Commercialising Fusion Energy

How small businesses are transforming big science

Commercialising Fusion Energy

How small businesses are transforming big science

Commercialising Fusion Energy

How small businesses are transforming big science

Edited by
William J Nuttall
School of Engineering and Innovation, The Open University, Milton Keynes, UK

Satoshi Konishi
Institute of Advanced Energy, Kyoto University, Gokasho, Uji, Kyoto, Japan

Shutaro Takeda
Urban Institute, Kyushu University, 744 Motooka, Nishi, Fukuoka, Japan

David Webbe-Wood
School of Engineering and Innovation, The Open University, Milton Keynes, UK

IOP Publishing, Bristol, UK

ISBN 978-0-7503-2719-0 (ebook)
ISBN 978-0-7503-2717-6 (print)
ISBN 978-0-7503-2720-6 (myPrint)
ISBN 978-0-7503-2718-3 (mobi)

DOI 10.1088/978-0-7503-2719-0

Version: 20201201

IOP ebooks

British Library Cataloguing-in-Publication Data: A catalogue record for this book is available from the British Library.

Published by IOP Publishing, wholly owned by The Institute of Physics, London

IOP Publishing, Temple Circus, Temple Way, Bristol, BS1 6HG, UK

US Office: IOP Publishing, Inc., 190 North Independence Mall West, Suite 601, Philadelphia, PA 19106, USA

In memory of Stephen N P Smith

In memory of Stephen N P Smith

Contents

Section IV Challenges and future opportunities

Preface

Editor biographies

William J Nuttall

Corresponding Editor William Nuttall CPhys FInstP is Professor of Energy at The Open University, UK. He is also a Fellow of the Payne Institute for Public Policy at the Colorado School of Mines, USA and a Fellow of the Construction Engineering Masters programme at the University of Cambridge, UK. He is an Affiliated Lecturer in Cambridge University Engineering Department and a Fellow of Hughes Hall, University of Cambridge. He holds a PhD in Physics (1993) from the Massachusetts Institute of Technology, USA, where he was a Fulbright scholar (1987–1988).

Satoshi Konishi

Editor Satoshi Konishi is a Professor at the Institute of Advanced Energy, Kyoto University. He was previously the Director of the Fusion Reactor Systems Laboratory at the Japan Atomic Energy Research Institute. He earned his PhD from the University of Tokyo.

Shutaro Takeda

Editor Shutaro Takeda is a Fellow of the Urban Institute, Kyushu University. He conducts research on energy sustainability, energy economy and renewable energy policy in developing nations. He is also the Research Representative of the Public Outreach Initiative for Nuclear Fusion at the National Institute for Fusion Science. In 2018, he was named the International Young Energy Professional of the Year by the Association of Energy Engineers.

David Webbe-Wood

Editor David Webbe-Wood is a PhD Student in the School of Engineering and Innovation at The Open University. He previously earned his Master of Research qualification in Nuclear Energy from Imperial College London. He is a member of The Society for Radiological Protection and a former Radiological Protection Specialist.

List of contributors

Ian Chapman
United Kingdom Atomic Energy Authority, Culham Centre for Fusion Energy, Culham Science Centre, Abingdon, Oxon OX14 3DB, UK

Tris Denton
United Kingdom Atomic Energy Authority, Culham Centre for Fusion Energy, Culham Science Centre, Abingdon, Oxon OX14 3DB, UK

Sehila M Gonzalez de Vicente
The International Atomic Energy Agency, Wagramer Strasse 5. A-1400 Vienna, Austria

Takuya Goto
National Institute for Fusion Science, 322-6 Oroshi, Toki, Gifu, Japan

Yoshita Mori
The Graduate School for the Creation of New Photonics Industries, Hamamatsu, Shizuoka, 431-1202, Japan

William Morris
United Kingdom Atomic Energy Authority, Culham Centre for Fusion Energy, Culham Science Centre, Abingdon, Oxon OX14 3DB, UK

Hoseok Nam
Graduate School of Energy Science, Kyoto University, Yoshidahonmachi, Sakyo, Kyoto: 606-8317, Japan

Bhavin Patel
United Kingdom Atomic Energy Authority, Culham Centre for Fusion Energy, Culham Science Centre, Abingdon, Oxon OX14 3DB, UK
York Plasma Institute, Department of Physics, University of York, Heslington, York YO10 5DD, UK

Richard J Pearson
Kyoto Fusioneering Ltd, 5-24 Gokashohirano, Uji, Kyoto, Japan

Garry Voss
United Kingdom Atomic Energy Authority, Culham Centre for Fusion Energy, Culham Science Centre, Abingdon, Oxon OX14 3DB, UK

Chris Waldon
United Kingdom Atomic Energy Authority, Culham Centre for Fusion Energy, Culham Science Centre, Abingdon, Oxon OX14 3DB, UK

Howard Wilson
United Kingdom Atomic Energy Authority, Culham Centre for Fusion Energy, Culham Science Centre, Abingdon, Oxon OX14 3DB, UK
York Plasma Institute, Department of Physics, University of York, Heslington, York YO10 5DD, UK

Melanie Windridge
Tokamak Energy Ltd. 173 Brook Drive, Milton Park, Oxfordshire, OX14 4SD, UK

Chapter 1

Introduction

William J Nuttall, Satoshi Konishi, Shutaro Takeda and David Webbe-Wood

1.1 Background

Nuclear fusion, the energy that makes the Sun shine, is regarded by many as the ultimate energy supply for humankind [1], and its promise of clean, safe, and virtually unlimited energy has driven global research and development for the last several decades. During this period, fusion research has sometimes met with criticism that the progress is too slow, or perhaps, stagnated. Today, however, there are indications that something is changing, the tide has turned. One can see that fusion energy makes headlines almost every day on major news outlets like Forbes[1] and Bloomberg[2]. The Economist, which ten years ago, observed 'Viable nuclear fusion has been only 30 years away since the idea was first mooted in the 1950s' [2] now reports with greater optimism: 'There is, then, no shortage of ideas about how a practical fusion reactor might be built.... Everyone talks a good story about this.' [3] This reinvigorated optimism was partly brought in by relatively new players in the fusion community: start-up companies.

Historically, fusion research has been the domain of large public institutions. These mainstream efforts seek to bring fusion to fruition as a source of energy and have been led initially by national governments and lately by international collaborations. The current international effort is focused on the ITER program—an international project with 36 nations participating directly or indirectly[3] at a cost of over €20 billion [4]. It is expected that the first deuterium-tritium operations of ITER will occur sometime after 2035.

[1] For example, 'Fusion Energy: Who Has The Courage To Take It To Market?' by Wal van Lierop on August 21, 2019.

[2] For example, 'Nuclear Fusion Could Rescue the Planet From Climate Catastrophe' by Jon Asmundsson and Will Wade on September 28, 2019.

[3] Members of the European Union are represented in this project via the Union. It is not clear at the time of writing what the position of the United Kingdom will be after Brexit.

doi:10.1088/978-0-7503-2719-0ch1

ITER will be followed by a set of DEMOs, an electricity-producing fusion power plant that is the last step to the commercial plants. It is anticipated that construction of DEMO, or set of DEMOs, will commence after 2040 [5, 6]. In this established sequence, the first demonstration fusion power plant would start supplying electricity around 2050 (see chapter 9). The start-up companies are striving to shortcut this path by 20 years using more innovative approaches. Some start-ups report on ambitions to start electricity generation even earlier, perhaps even as early as by 2025 (see chapters 5 and 7).

To illustrate how the landscape is changing in the global fusion community, the authors calculated the percentages of private versus public effort in terms of the number of devices[4] as shown in figure 1.1. This comparison shows that while only 10% of the currently operating fusion devices are owned by the private sector, half of

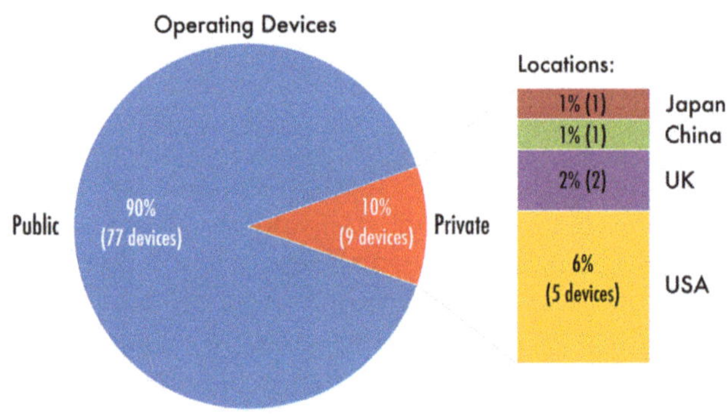

(a) Operating Devices

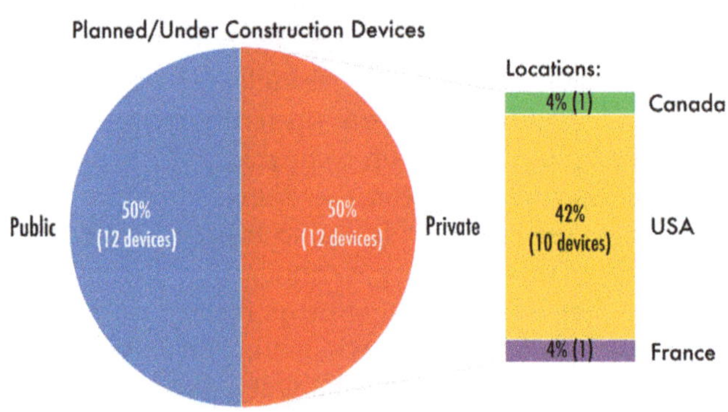

(b) Planned or Under Construction Devices

Figure 1.1. Percentages of fusion devices in the world in 2020: public versus private (based on [7]). (a) Operating devices (b) planned or under construction devices.

[4] Based on the IAEA Fusion Device Information System (FusDIS), a directory of fusion experimental facilities worldwide available online on https://nucleus.iaea.org/sites/fusionportal/Pages/FusDIS.aspx.

the planned or under construction fusion devices are already led by private entities. In fact, in the United States, the majority of fusion devices now belong to the private sector.

The rise of private fusion enterprises would have been inconceivable even a decade ago. Fusion research was once looked upon as being a synonym of 'big science'—*large-scale scientific research consisting of projects funded usually by a national government or group of governments*[5]. While it might still be appropriate to assess this global trend cautiously, such enterprises are gaining more and more support from investors. Prominent investors like Jeff Bezos and Bill Gates are among the supporters of this emerging sector. As a result, there are at least 40 private enterprises pursuing faster commercialization of fusion globally, with around $2 billion raised in total as of 2020 (chapter 7). These forward-thinking start-ups are competing to be the first to deliver on the long-overdue promise of virtually unlimited clean energy: they are transforming *big science*.

1.2 What is nuclear fusion?

But what is fusion? At its simplest, nuclear fusion is the application of Einstein's famous equation:

$$E = \Delta mc^2$$

When two light nuclei are fused together to form a heavier nucleus[6] (sometimes releasing other particles), the mass of the resulting nucleus and emitted particles slightly less than that of the two initial nuclei. This 'lost' mass is released as kinetic energy, as illustrated in figure 1.2. This is nuclear fusion.

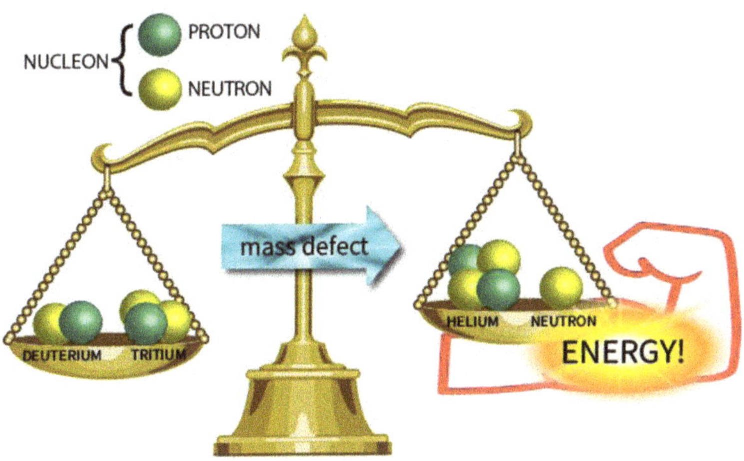

Figure 1.2. Fusion converts tiny amounts of mass into vast amounts of energy. Courtesy of Aya Kuretani.

[5] Marriam-Webster: https://www.merriam-webster.com/dictionary/big%20science

[6] In some cases, more than one nucleus is produced

Nuclear fusion is responsible for the production of energy in stars. At the core of stars, protons react with other protons to form deuterium nuclei[7] (^2H) and positrons. The deuterium nuclei can merge to form helium-4 nuclei (^4He), or they can interact with other protons to form helium-3 (^3He). Two helium-3 nuclei can fuse to form a nucleus of an unstable beryllium isotope (^6Be) that breaks apart to give ^4He and two protons. The energy released at each step causes stars to shine, including the Sun. As such, fusion can be regarded as the ultimate source of energy: life on earth would not be possible without fusion reaction.

In addition to the fusion processes briefly outlined above, a number of other fusion reactions exist. Examples include (where Q is the energy released in the reaction):

$^2H + {}^2H \rightarrow {}^3T + {}^1H$, $Q = 4.04$ MeV;
$^2H + {}^2H \rightarrow {}^3He + n$, $Q = 3.27$ MeV;
$^2H + {}^3H \rightarrow {}^4He + n$, $Q = 17.6$ MeV;
$^1H + {}^{11}B \rightarrow 3({}^4He)$; $Q = 8.68$ MeV;
$^1H + {}^6Li \rightarrow {}^3He + {}^4He$; $Q = 4.023$ MeV;
$^3He + {}^3He \rightarrow {}^4He + 2p$; $Q = 12.9$ MeV;
$^3He + {}^6Li \rightarrow {}^1H + 2({}^4He)$; $Q = 16.88$ MeV;
$^3He + {}^6Li \rightarrow {}^2H + {}^7Be$; $Q = 0.113$ MeV.

The reaction of principal interest to most commercial fusion developers is the deuterium (^2H or D)—tritium (^3H or T) reaction, as this 'D-T reaction' is regarded as the easiest to achieve [8]. However, a small number of companies, including TAE of the United States[8], are developing concepts based on other reactions (see chapter 7).

Unfortunately, the nuclei carry a positive electrostatic charge and repel each other before they can fuse. For fusion to occur, the plasma containing the fusion fuels must reach the thermal (kinetic) energy required, which requires both the containment and the heating of the plasma.

To use fusion as an energy source, the energy released by fusion reaction must be greater than the energy that is required to induce the reaction. The ratio of the energy output from nuclear fusion reactions in the plasma to the energy supplied to sustain the plasma is known as the fusion energy gain Q_{fus} [8]. $Q_{fus} = 1$ is referred to as the break-even condition.

Q_{fus} is closely linked to the plasma density n, the plasma temperature T, and the efficiency of contained thermal energy (confinement time τ_E). The combination of these three factors $nT\tau_E$ is known as the Lawson criterion, which is used to evaluate the performance of a fusion reactor. A higher performance for this combination of factors can be achieved in different ways, and this is where the ingenuities of the scientists are put to the real test. In magnetic confinement approaches, magnetic fields are applied to increase the confinement time, whereas in inertial confinement

[7] The ^2H species is often referred to a deuterium and the symbol D used. Similarly, the ^3H nuclei are referred to as tritium with the symbol T.
[8] Formerly known as Tri Alpha Energy

approaches, use is made of lasers or other techniques to compress targets to provide the high densities required. Both approaches are described in chapter 2.

When realized, fusion energy may have significant advantages over other low carbon energy sources: it does not suffer from the issues of variability that affect wind and solar; compared to nuclear fission, the issues of radioactive waste are much reduced. In addition, there are no risks of runaway accident; and above all, since the deuterium fuel can be extracted from seawater and lithium remains an abundant metal[9], fusion energy has almost unlimited fuel resources on earth—it is estimated that fusion could sustain humanity for the next one billion years [9]. That said, fusion is not without resource sustainability concerns, for example as regards helium as discussed in Chapter 11, section 11.7.1.

1.3 Purpose and structure of this book

This book aims to give readers, with or without a background in nuclear science or engineering, an appreciation of where the state of the art is and the future prospects for commercialization. This book is, in part, a follow up of a workshop organized jointly by the Open University (UK) and Kyoto University (Japan) under the auspices of the Anglo Japanese JUNO network (JUNO: A Network for Japan-UK Nuclear Opportunities) at Hughes Hall, Cambridge University in June 2019.

The book is organized into four sections. The first section, *Private Fusion Primer*, gives a brief overview of private fusion. In chapter 2, Dr Richard Pearson from Kyoto Fusioneering Ltd and Dr Shutaro Takeda from Kyushu University introduce the basics of fusion reactor engineering to the readers. Next, Dr Shutaro Takeda and Dr Sehila M Gozalez de Vicente from the International Atomic Energy Agency (IAEA) provide considerations for the commercialization of the technology in chapter 3. Subsequently, in chapter 4, Mr David Webbe-Wood summarizes possible funding schemes for future commercial fusion power plants.

The second section, *Progress in the Private Sector*, showcases the latest progress of the private fusion start-ups. First, Dr Melanie Windridge from Tokamak Energy Ltd. presents an argument for their approach, progress, and plans in chapter 5. Subsequently, in chapter 6, Professor Yoshitaka Mori from The Graduate School for the Creation of New Photonics Industries shares with the readers their public–private CANDY laser fusion concept. Chapter 7 provides an overview of the approaches of other private fusion companies by Dr Richard Pearson and Professor William Nuttall.

The third section, *Public Sector Push to Commercialization*, outlines the public sector's efforts toward commercialization. Professor Howard Wilson and his colleagues from the UK Atomic Energy Authority describes the UK's STEP programme in chapter 8. Next, in chapter 9, Professor Takuya Goto from National Institute for Fusion Science (NIFS) presents a technical overview of the Japanese public strategy for magnetically confined fusion energy.

[9] That is, notwithstanding the pressures placed on its supply by the burgeoning battery manufacturing industry.

The fourth section, *Challenges and Future Opportunities*, discusses the challenges and future opportunities of private fusion. In chapter 10, Dr Richard Pearson gives general observations on the challenges private fusion companies have to overcome. Professor William Nuttall provides the historical background of the private fusion and extrapolates the story to future opportunities in chapter 11. Professor Satoshi Konishi and Dr Shutaro Takeda then expand the idea to utilize nuclear fusion as a means of decarbonization in chapter 12. Finally, the book ends with the conclusions and thoughts of the editors.

Only time will tell when and how humanity realizes commercial fusion energy. It seems that the second decade of the 21st century represents the key moment that will determine the future of a concept that has been imagined for many decades. As the chapters that follow will discuss, the public sector efforts thus far have been essential in making the private sector initiatives possible. Whether the leadership in the field passes from the public to the private sector, or whether the relationship will be closer and more collaborative than that, will be something that we will see emerge in the coming years. What seems clear, however, is that fusion faces its best opportunity in many decades to find a bold and ambitious way ahead.

References

[1] Dean S O 2013 *Search for the Ultimate Energy Source: A History of the US Fusion Energy Program* (New York: Springer)

[2] The Economist 2010 Funding for fusion an expensive iteration *The Economist*

[3] The Economist 2019 Doughnuts, apples, smoke rings *The Economist*

[4] Claessens M 2020 *ITER: The Giant Fusion Reactor* (New York: Springer)

[5] Nordlund K H 2018 European research roadmap to the realisation of fusion energy (Munich: EUROfusion) www.euro-fusion.org/eurofusion/roadmap

[6] Okano K, Kasada R, Ikebe Y, Ishii Y, Oba K and Kashiwagi M *et al* 2018 An action plan of Japan toward development of demo reactor *Fusion Eng. Des.* **136** 183–9

[7] International Atomic Energy Agency 2020 The fusion device information system (FusDIS) (https://nucleus.iaea.org/sites/fusionportal/Pages/FusDIS.aspx)

[8] Takeda S and Pearson R 2019 *Nuclear Fusion Power Plants. Power Plants in the Industry* (London: IntechOpen) p 101–21

[9] Bradshaw A M, Hamacher T and Fischer U 2011 Is nuclear fusion a sustainable energy form? *Fusion Eng. Des.* **86** 2770–3

Section I

Private fusion primer

Chapter 2

Review of approaches to fusion energy

Richard J Pearson and Shutaro Takeda

2.1 Introduction

In recent years, private industry has started to become more involved in the fusion research space; a space previously occupied almost solely by programmes supported by government funding [1]. A surge in new interest, mostly in North America, has resulted in more than 20 private fusion companies having been established, to date (see chapter 7). As summarized in the Introduction of this book, at least nine research devices are in operation by private fusion companies worldwide, while another 12 are under construction or are planned [2, 3]. In context, this means that around 20% of all fusion research devices are now owned by the private sector. The private sector's approach to fusion energy is worth specific attention—as will be explored in several chapters of this book—as many are pursuing reactor designs that reflect 'outside of the box' thinking. In fact, of the 21 devices, 15 are classified as 'Innovative/Alternative Concepts' under the IAEA's fusion reactor classification criteria, meaning they do not follow conventional designs. Figure 2.1 compares the percentage of designs being developed by public fusion programmes and by the private sector. The comparison clearly illustrates the more innovative and enterprising nature of the private sector: the percentage of 'Innovative/Alternative Concepts' is almost four times larger in the private sector (71% versus 19%). Evidently, tokamaks, which have formed the backbone of publicly funded fusion programmes for over 50 years, are not the primary focus of private sector development efforts[1].

The range of fusion reactors that are currently under development, which will be detailed and referred to throughout this book, fall into one of three broad conceptual spaces. These are magnetic confinement fusion (MCF), inertial

[1] Despite the share of private companies focused on tokamaks being only 14% in terms of the number of devices, these companies have been awarded around one quarter of all private sector investment (see chapter 7).

doi:10.1088/978-0-7503-2719-0ch2

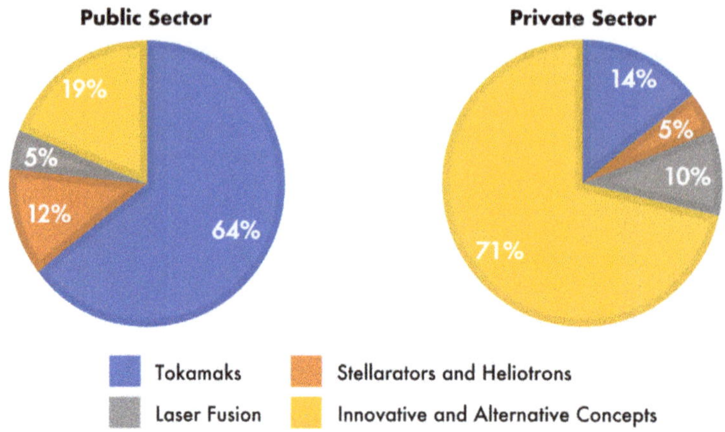

Figure 2.1. Comparison of design approaches: private versus public, produced using data from [3].

confinement fusion (ICF), and magneto-inertial fusion (MIF)[2]. Each of these spaces can broadly be defined by the density of the fusion plasma. MCF reactors and ICF reactors have low- and high-density plasmas, respectively, and MIF reactors have plasma densities spanning the region between MCF and ICF. There are multiple reactor variants within each of these spaces. As shown in figure 2.1, and described in more detail later in chapter 7, it is in fact in the MIF space that most private-sector fusion companies have emerged. This is likely because public sector programmes have historically tended towards the exploration and development of MCF and ICF approaches, rather than MIF which, by comparison, is largely unexplored [4]. The following list of concepts, all of which sound straight out of a science-fiction movie, represent just several of the reactor designs under development: beam-target, dense plasma focus, field reversed configuration, inertial electrostatic, levitated dipole, magnetic mirror machine, magnetized target, Z-pinch, pulsed power generated electron beam, reversed field pinch, simple magnetized torus, and the spheromak.

The aim of the first half of this chapter is to review the approaches to fusion, in order to provide a foundation level understanding for the reader to better traverse the later chapters of this book. As such, it will first outline the three broad conceptual spaces—MCF, ICF and MIF—to provide a fundamental understanding of the approaches and how they differ[3]. As will become clear, despite having the same goal, which is broadly to develop a reactor capable of demonstrating a power-producing fusion reaction, reactor designs—even within each conceptual space—are quite different indeed. Despite key differences, however, there are commonalities in physics and engineering. Accordingly, the second half of this chapter provides an assessment of the core engineering challenges associated with developing a functioning fusion reactors, focusing on each of the three conceptual spaces in turn.

[2] MIF is sometimes referred to as magnetized target fusion (MTF), but here MTF is taken to be a specific subset of MIF.

[3] Details of specific concepts within these three spaces are provided in the chapters in the latter half of this book, two of which that are contributed by private sector companies (see chapters 5 and 6).

The chapter will refer to specific reactor concepts, but only to aid fundamental understanding. Specific reactor concepts will be described in greater detail in the later chapters of this book.

2.2 Three key conceptual spaces for fusion reactors: MCF, ICF and MIF

Before the three conceptual spaces can be detailed, it is first instructive to understand that the key criterion for all fusion reactors which is important irrespective of the conceptual approach, known as the fusion gain (Q_{fus}). Fusion gain is the ratio of power produced by a fusion reaction to the power absorbed by a plasma in a reactor. The goal for all fusion developers is to create a reactor capable of a Q_{fus} greater than one; in other words, where the power produced is greater than the power required[4]. Q_{fus} depends on the fusion triple product, which is the product of plasma temperature, plasma density, and confinement time (how long the fusion reaction in a plasma is sustained). The values of these three factors, in particular the density (as noted in the previous section), is different for each of the three conceptual spaces[5].

MCF approaches use magnetic fields to confine and manipulate fusion reactions in a steady-state or long-pulse plasma. These plasmas are typically relatively low-density, which means it is necessary to have a longer confinement time. MCF reactors also require the plasma to be heated to increase its temperature. This is usually provided by way of external beams or electromagnetic waves targeted at the plasma, which brings it to the conditions necessary to achieve a sufficiently high fusion triple product. ICF, at the other end of the spectrum, strives to generate a plasma of extremely high density and temperature, but with a very short confinement time. Therefore, ICF reactors are inherently pulsed machines. MIF reactors combine aspects of MCF and ICF. Compared to MCF, they create relatively high-density plasmas but with relatively short confinement times (though not as short as that of ICF). The medium density plasma is then compressed by a driver, not unlike in an ICF reactor to achieve a sufficiently high fusion triple product. With this fundamental understanding, each of these approaches can now be explored in more detail.

2.2.1 Magnetic confinement fusion

Plasmas are positively charged and can be manipulated by magnetic and electrical fields. In an MCF fusion reactor, sets of magnets are used to create and confine a fusion plasma. Tokamaks are by far the most researched and thus most common type of fusion reactor—not just within the MCF space, but across all the conceptual

[4] For commercial fusion energy to be viable, the fusion gain must be significantly higher than one. For example, a Q_{fus} of 10 may yield an overall system gain of only ~1, after losses in energy conversion and overall site power are considered. This is commonly referred to as 'engineering gain', or Q_{eng}.

[5] There are several other approaches to fusion, including some more abstract subsets of MCF, ICF and MTF fusion, that are not explicitly detailed. The challenges detailed in this chapter do not necessarily apply to, for instance, inertial electrostatic confinement fusion approaches.

spaces. In a tokamak, a set of toroidal magnetic coils confines a ring plasma containing fusion fuel (typically the heavy isotopes of hydrogen, deuterium and tritium). A central solenoid magnet at the centre of the ring plasma is then used to drive current in the confined plasma to produce a helical magnetic field. Most tokamaks use a configuration that produces this donut-shaped plasma, similar to that shown in Figure 2.2.

The pathway to fusion based on the tokamak has been explored by the public sector for the past few decades. Just a few key tokamak reactors are the *Tokamak Fusion Test Reactor* (TFTR) in Princeton, USA (which started operation in 1982 and ended in 1997); *Joint European Torus* (*JET*) in Culham, UK (which started operation in 1983); *Aditya* in Gujarat, India (1989); *EAST* in Hefei, China (2006); *KSTAR* in Daejon, Korea (2008); and *JT-60SA* in Naka, Japan (1985, upgraded in 2020). Despite being the most extensively researched reactor, with over 200 tokamaks having been built over the past 60 years or so, not one has yet demonstrated a Q_{fus} of above one. Gradual progress towards net fusion gain towards the later part of the 20th century provided enough understanding of tokamak physics and engineering for *ITER*—an international project to build a tokamak that would provide the first demonstration of the goal that had thus far eluded fusion scientists—to be conceived. Constructed in the South of France and supported by several of the world's nations, including Japan, the EU and Russia, ITER will begin operation in 2025. The ITER project is buttressed by several tokamak experiments, including campaigns on JET and JT-60SA. In the private sector, two companies are also pursuing tokamaks based on a similar design to

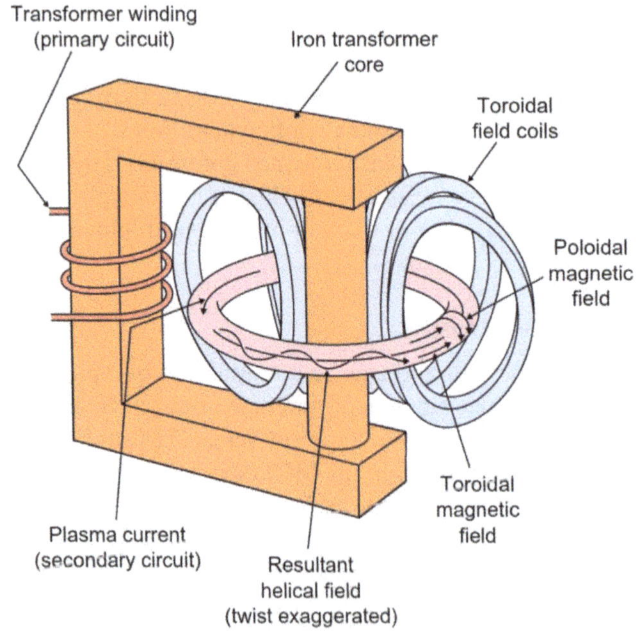

Figure 2.2. Schematic of a tokamak-type MCF reactor. Figure reproduced from [5], copyright 2005, with permission from Elsevier.

ITER. Whilst General Atomics, which operates a user facility in San Diego, California (US) can be considered a private organization very much involved in fusion, the company does not—as far as the authors are aware—intend to develop a commercial fusion reactor. By contrast, the more recent emergence of Commonwealth Fusion Systems, an MIT spin-off, is an example of a private fusion company pursuing a conventional tokamak pathway to commercial fusion energy (see chapter 7).

An additional approach to fusion is via the spherical tokamak, a reactor configuration very similar to the tokamak, but with a lower aspect ratio (the ratio of the outer to the inner radius of the plasma) and a higher plasma elongation that results in a plasma shaped more like a cored apple rather than a donut. Several developers in the public sector are pursuing spherical tokamaks, based on the success of devices like MAST in Culham, UK, and NSTX at Princeton Plasma Physics Laboratory, USA. The STEP programme, also at Culham, UK (see chapter 8), as well as at Princeton Plasma Physics Laboratory in the US, are both coordinating spherical tokamak programmes. Additionally, one private company, Tokamak Energy Ltd in Oxford, UK, is pursuing a pathway via the spherical tokamak (see chapter 5).

Whilst tokamak physics is regarded as the most well understood as regards fusion physics, particularly compared with approaches in the other two conceptual spaces, there remain several unsolved problems. One such problem is how to prevent a plasma disruption—an extremely fast event that is hard to predict, which can result in the complete loss of stored energy in the plasma. Another problem is the understanding of a burning plasma when significant alpha heating occurs as a result of the alpha particles produced by the D–T fusion reaction transferring their heat to the plasma—a never-seen-before phenomenon which is expected to occur at a Q_{fus} value of approximately 3.

Stellarators are another type of MCF reactor. Like tokamaks, they use magnetic coils. However, the coils on a stellarator are shaped into complex geometries to create a magnetic field that both confines the plasma and drives current, eliminating the need for a central solenoid and separate toroidal and poloidal field coils. While it is more challenging to build such magnets, unlike tokamaks, stellarators benefit from inherent plasma stability and do not require systems to drive current, such as gyrotrons [5, 6]. Like tokamaks, the stellarator pathway to fusion is being explored in various publicly funded research programmes across the world. The *Wendelstein 7-X* in Germany, *LHD* and *Heliotron-J* in Japan, *TJ-II* in Spain, and *HSX* in the USA have all made particularly significant contributions to the understanding of stellarators in recent years, in particular through a strong collaboration between the research groups under the IEA Technology Collaboration Programme on Stellarator-Heliotrons. In addition to publicly funded programmes, a private company very recently launched in France, Renaissance Fusion, to develop the first privately funded Stellarator reactor with the mission of realizing fusion energy (see chapter 7).

Whilst other MCF reactor types have been built and tested—including the field-reverse configuration (FRC) being explored by TAE Technologies in the US, described later in chapter 7—tokamaks and stellarators are by far the most common type of MCF reactor under development.

A list of MCF devices in the world are assembled based on IAEA's FusDIS [3] as a supplementary table at the end of the chapter (table 2.S1).

2.2.2 Inertial confinement fusion (ICF)

ICF reactors seek to create the conditions for fusion by instantaneously and rapidly compressing and heating a small fuel target by using one or more high-power lasers (or beams). As such, ICF is based on the fundamental principle of the hydrogen bomb, whereby the fuel in the target is compressed and heated quickly to reach the conditions to burn before energy confinement is lost. Of course, the key difference to an H-bomb is that in an ICF reactor the reaction is carried out in a controlled and repeatable manner [5][6]. As noted previously, ICF reactors are pulsed systems. Inter alia, to generate a steady output of energy, ICF reactors require both fuel target injection and the laser system to be highly repeatable, and for the reactor chamber to withstand the repeated force from continuous pulsed implosions of the fuel target.

Several ICF configurations are being explored. The two most common approaches, indirect and direct drive, are shown in figure 2.3. In indirect drive, a fuel target is positioned inside a metal cavity called a hohlraum, which is typically made of a heavy metal such as lead or gold. Lasers are targeted at the hohlraum, rather than at the fuel target itself, to generate x-rays which then heat and compress the fuel target to the conditions for fusion to occur [8]. Whilst some of the laser energy is lost in the conversion to x-rays, the energy is more uniformly absorbed by the fuel target leading to a more stable and symmetrical implosion. A small fraction of the fuel target, the 'hot spot' in the centre and the surrounding primary fuel region of the target, is then heated and compressed and the reaction propagates outwards, achieving a high fusion triple product by way of very high density and temperature but with a very short confinement time [5]. The national ignition facility (NIF) in the

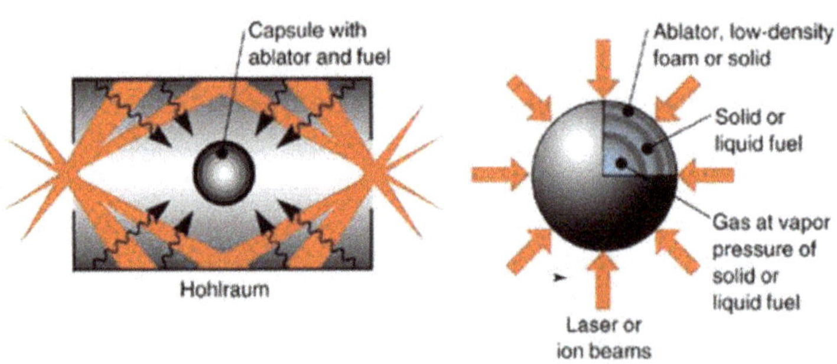

Figure 2.3. Mechanisms of ICF—indirect drive (left) and direct drive (right). Figure reproduced from [7] with permission. Note that the original figure has been cropped.

[6] Consequently, for its function to understand thermonuclear fusion reactions, much ICF research has been supported under the auspices of national defence and security budgets rather than energy budgets except in Japan [15].

USA—which has suffered several performance setbacks that throws the whole conceptual approach of indirect fusion into question[7]—is researching an indirect drive ICF via the use of a hohlraum targeted by 192 lasers.

In direct drive ICF, the fuel target is directly irradiated by the laser or ion beam. Direct drive ICF allows for higher efficiency in converting laser energy into energy striking the fuel target compared with indirect drive. However, difficulties in uniformly ablating the fuel target results in the rapid growth of nonuniformities, which results in reduced compression and break-up of the fuel target before it reaches a sufficiently high density and temperature for fusion [5]. For direct drive, the accuracy and the power profile of the laser—that is, the amount of power delivered over the time of the pulse—must be coupled to the design of the fuel target. Specifically, the fuel target must be fabricated such that the outer layers are ablated and compressed before fusion conditions are achieved in the centre. The *Omega facility* at Laboratory for Laser Energetics at the University of Rochester, USA, and the *GEKKO XII* laser facility at Osaka University, Japan, are both exploring direct-drive ICF [8].

Alternative ICF configurations include fast ignition, in which the compression and heating stages of the process are separated; a primary high-power laser first compresses the fuel then an ultra-intense-high-power secondary laser heats the core [5]. The Japanese *FIREX* projects and other fast ignition programmes are ongoing to improve coupling efficiency from ultra-high intense laser energy with fast electrons or fast ions [9–13]. Progress on such laser-based ICF approaches is detailed later in this book (see chapter 6).

Finally, one more novel ICF approach is that of First Light Fusion Ltd, a privately funded company in the UK, which is exploring a reactor that does not use lasers at all and is instead based on 'shock ignition' [14]. Shock ignition is similar to fast ignition, whereby a fuel target is first compressed to ultra-high density and then imploded by a high-intensity shockwave [5, 15]. The First Light Fusion approach is described in more detail in chapter 7, but the shock ignition concept is mentioned here as it provides a useful bridge between the ICF and MIF space.

A list of ICF devices in the world currently are assembled based on IAEA's FusDIS [3] as a supplementary table at the end of the chapter (table 2.S2).

2.2.3 Magneto-inertial fusion (MIF)

As the name 'magneto-inertial' suggests, MIF approaches operate in the parameter space between the low-density MCF and high-density ICF conceptual spaces, and duly incorporates aspects of both [16]. In MIF reactors, a target, typically—but not always in the form of a self-contained plasma (a 'plasmoid') is created and injected into the centre of the MIF reactor chamber. This target is then compressed and

[7] The US NIF experiment (on the 'National Ignition Campaign') was concluded in 2012 when, despite modeling carried out beforehand showing high performance, experiments only yielded a fusion reaction with around 10% of the conditions required for ignition (the point at which energy released from an ICF fusion reaction is greater than the losses) [15]. Experiments are ongoing to identify and address phenomena that affect implosion quality in indirect-drive ICF [45], but considerable theoretical and practical work is needed for the concept to be proven as a viable pathway to fusion energy.

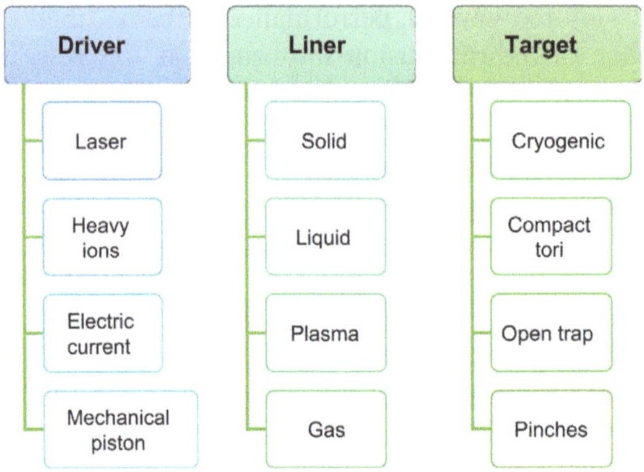

Figure 2.4. The range of possible drivers, liners, and targets for MIF reactors, based on [17]. The left-hand column shows the types of driver (which determines how energy is delivered to the liner), the central column shows the types of liner (which is the medium to provide compression to the target), and the right-hand column shows the types of target (containing the fusion fuel, most commonly in the form of a plasma). The examples of driver, liner and target are not exhaustive. Adapted from [7] with permission of Pleiades Publishing.

heated by a liner (a solid, liquid or gas that surrounds the fuel target), which is acted upon (or in some cases propelled) by a driver, to conditions of high enough density and temperature for fusion. Accordingly, MIF reactors are often referred to—sometimes interchangeably—as 'magnetized-target fusion' (MTF) reactors[8], because most involve the creation of a plasma that acts as a target to be struck by a liner that compresses it. MIF reactors are—like ICF reactors—pulsed systems, and the sequence of creating and injecting, and then compressing and heating a plasma is recurring. The diagram in figure 2.4 shows the three primary components—the driver, liner and target—in a MIF reactor. Different MIF concepts combine one component from each of the three columns. Figure 2.5 is a schematic of the fuelling and exhaust systems, adapted from similar diagrams for ICF, to show the basic configuration of both pulsed MIF and ICF reactors.

Examples of MIF reactors include MagLIF, a US project, which blurs the lines with ICF by combining the use of lasers to achieve compression whilst heating the target to produce a magnetized plasma [8, 18]. MagLIF thereby necessitates developments in both laser and magnet technology. Another example of an MIF reactor is that of General Fusion, a privately-funded Canadian company, which employs mechanical pistons (powered by compressed gas) as a driver, to compress a liquid liner onto a magnetized compact (spherical) plasma. A notably different MIF approach to that of General Fusion is the *Plasma-Liner Experiment* (PLX) which is being collaboratively developed between Los Alamos National Laboratory and private company HyperV Technologies Corporation. The PLX works by launching an array of high-speed plasma jets containing D–T fuel towards the centre of a spherical reactor vessel to merge and

[8] MTF is here considered as a specific subset of MIF, in the same way that tokamak approaches (of which there are several) are a subset of MCF.

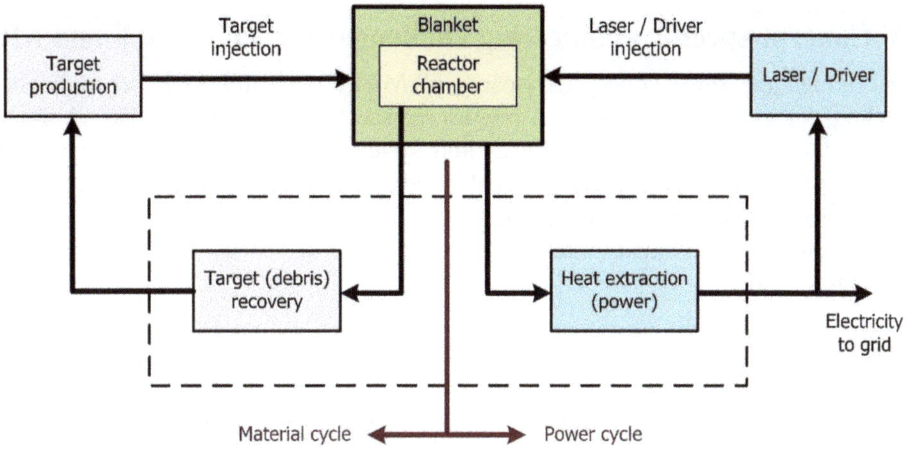

Figure 2.5. Reactor schematic for the fuelling and exhaust systems in ICF and MIF reactors, based on diagrams showing these systems specifically for ICF reactors in [28, 29]. Adapted from [28], copyright 2019, with permission from Elsevier.

form a plasma liner, which converges and implodes in the centre of the reactor chamber, where it is acted upon by a laser pulse to drive it to fusion conditions [19].

Arguments that MIF can avoid the problems associated with MCF and ICF while exploiting the benefits of each are mainly predicated upon the potential to simplify driver technology compared with ICF lasers; to avoid the need for the plasma to be as hot or stable for as long as in an MCF reactor; and to avoid the design complexity of both [20]. Furthermore, a MIF plasma does not need to be as hot, dense, or stable for as long as an MCF plasma, as it is heated and compressed by the liner. At the same time, the magnetic field in a MIF plasma can substantially reduce the flow of heat out of the fusion plasma compared with an ICF plasma, meaning that, a MIF reactor does not require compression as intense or as precise in an ICF reactor [16, 21]. It is for these reasons that proponents—mostly privately funded companies—are pursuing MIF. However, as shown by the various possible combinations in figure 2.4, and in the examples above, the type of reactor and potential benefits of avoiding MCF and ICF are dependent on the specific reactor concept. Additionally, despite progress in recent years, many in the fusion community deem MIF concepts to still be in their infancy compared to MCF and ICF. The approach by General Fusion and in the PLX, for example, have not been extensively researched or tested to the extent of tokamaks. MIF reactors are less developed and certainly less well understood in terms of the understanding of physics than both MCF and ICF. As such, fundamental physics understanding, akin to that provided by early tokamaks and stellarators, must be demonstrated before any MIF concept can be deemed as viable compared with these already established reactor designs, let alone for commercial fusion energy. Nonetheless, proponents of MIF hope that the perceived benefits can be demonstrated and that their concepts can leapfrog the currently slow progress in MCF, to become the frontrunner in the mission towards commercial fusion energy within the next decade [2].

A list of MCF devices in the world are assembled based on IAEA's FusDIS [3] as a supplementary table at the end of the chapter (table 2.S3).

2.3 Concept-specific engineering challenges for MCF, ICF and MIF

Even though the three conceptual spaces of MCF, ICF and MIF are different—including, most certainly, from a physics perspective—the engineering challenges associated with developing technologies and systems necessary to realize a working reactor can be grouped into three broad categories—all of which are, to an extent, linked:

- Plasma production and control
- Fuelling and exhaust systems
- Diagnostics

This section attempts to provide a fresh perspective of these engineering challenges by taking a 'concept neutral' view. A description of the challenges associated with each of the three categories within each conceptual space is provided, and reference to specific MCF, ICF and MIF reactor designs is made to underline differences across concepts. Note that the challenges within these three categories—which together make up the key systems in an experimental fusion reactor that is capable of demonstrating a fusion reaction that produces more power than that which is supplied to it—are the primary focus of all fusion developers. Indeed, the development of solutions to the challenges in these three broad categories should thus be seen as central to the success of all such developers. In the case of private companies, in particular, there is significant secrecy in disclosing full details of solutions to these challenges. Furthermore, whilst there is little potential for cross-cutting development across the conceptual spaces, where overlaps do exist they are detailed. There are also specific technology challenges that are shared not only within a conceptual space but also across the conceptual spaces. Chapter 10 later describes the broader and more cross-cutting engineering challenges which must be tackled by all fusion developers, regardless of the concept (MCF, ICF or MIF), if they are to move from a net-power producing fusion reactor to a first commercial fusion power plant.

2.3.1 Plasma production and control

It is in the production and control of the plasma that we find perhaps the greatest difference between MCF, ICF and MIF reactors. All approaches require different technologies to be developed. For MCF reactors, the key focus is on developing effective magnets to confine the plasma and on the systems required to drive current and heat the plasma to high enough temperatures for fusion. In ICF, it is on developing high-power drivers—predominantly, lasers—and also on the design of targets that can be imploded with high accuracy and high repeatability by those lasers. In MIF, it is on developing systems to create and inject a (plasma) target into the reactor chamber, as well as designing a coupled driver-liner system to compress that target. Note that while the challenges associated with engineering the target and the driver for a MIF reactor appear similar to the challenges for MCF and ICF, respectively, they are largely unique to the MIF approach.

First, let us explore the challenges for MCF more closely. There are many complex physics challenges associated with the development of MCF plasmas, which are explained in [22] and further characterized with regards to plasma shape, size, and field in tokamaks versus spherical tokamaks in [23]. However, here focusing on the engineering issues, we are most concerned with how to develop a strong and stable magnetic field to confine the plasma, as well as the means to heat the plasma to the high temperatures required for fusion. These two challenges are inherently coupled. The electric current in a fusion plasma causes ohmic heating. As a plasma gets hotter, electrical resistance falls, and the current becomes less effective at heating. Increasing the plasma current can increase the temperature of the plasma, but can also lead to potentially damaging plasma disruptions unless the magnetic field can be increased in line with the current. The density and energy confinement time of the plasma—two of the critical factors in the triple product—also increases with magnetic field. As such, the challenge is to increase the magnetic field to improve the plasma performance.

While controlling a plasma with high-power magnets can be considered as a physics challenge, the development of the magnet itself is fundamentally an engineering one. Up until the early 2000s, copper magnets were used in MCF reactors, when LTS (low-temperature superconducting) magnets were then developed. LTS magnets, which will be used in ITER (specifically, ITER's LTS magnets will be made of niobium-tin), can deliver much higher fields than copper magnets. However, LTS magnets must be cooled to extremely low temperatures (typically 4 K) to operate at such high fields. A further issue with LTS, in addition to these extreme cooling requirements, is the need for magnet quench protection. A magnet quench occurs when a magnet instantaneously loses superconductivity and dissipates its stored energy. This stored energy can destroy the magnet. To avoid the pitfalls associated with LTS magnets, as well as to enhance plasma stability, increase plasma current and improve overall efficiency, there has been great interest in developing high-temperature superconducting (HTS) magnets. HTS magnets can operate at a higher field and current density than LTS magnets. They are more efficient due to their comparably higher operating temperature—typically around 20 K, rather than 4 K—reducing the need for powerful cryogenic systems and substantial thermal shielding (to keep the magnets at operating temperature) [23, 24]. HTS magnets were initially explored in public sector programmes, including at KIT, Germany, with the Roebel conductor, at MIT, USA, with the TSTC conductor, at Tohoku University, Japan, and also in a dedicated programme at NIFS, also in Japan. These research groups continue to conduct intensive activities including the organization of HTS4Fusion Conductor Workshop. In the past decade or so, however, the most significant HTS magnet development has been in the private sector. In particular, this has been for tokamaks at Tokamak Energy Ltd in the UK and Commonwealth Fusion Systems in the USA, and for the stellarator under development at Renaissance Fusion in Spain. Key engineering challenges for HTS magnet development are demonstrating stable operation at high-field and high-current, as well as the electromechanical design and manufacturing of HTS tape [23]. Further details on the challenges associated with developing HTS magnets are provided in chapter 5, contributed by Tokamak Energy Ltd

While increasing the magnetic field and thus increasing the current may increase plasma temperature, additional systems to heat the plasma are still required. Two methods for providing additional heating are most commonly considered: neutral beam injectors (NBIs), which introduce high-energy particles to the plasma via a beam, and high-frequency gyrotrons, which use microwaves to heat the plasma by thermal resonance. NBIs are being installed on ITER and are being considered for other MCF reactor experiments [25]. However, the power required for NBIs to heat a plasma to fusion conditions is significant, so much so that it would make up a substantial portion of the recirculating power for a fusion power plant; that is, NBIs are likely to reduce overall plant efficiency [26]. High-frequency gyrotrons are well established and, unlike NBIs, do not require interfacing directly with the MCF reactor vessel. However, if they are to be used, for instance, with tokamaks with higher fields due to HTS magnets, the frequency and power of gyrotron heating must be increased if they are to be a viable solution for plasma heating, as the required frequency of the gyrotron is approximately proportional to the magnetic field.

Unlike in MCF reactors, the control of an ICF plasma relates instead to the striking and ignition of the fuel target. The control mechanism for an ICF reactor has to be such that it allows repeatable shots over very short periods, and with continuous high accuracy. This requires a laser to deliver power accurately and repeatedly to the fuel target, and the design of the fuel target to be such that it is compressed symmetrically when struck by the incident laser [27].

Both indirect and direct drive approaches require the energy profile of the laser—again, the amount of power delivered over the duration the laser pulse—to allow uniform compression and heating of the fuel target. Ideally, the outer layer of the target should be ablated with low intensity, and then the laser power should increase to a maximum when reaching the centre of the capsule. The geometry and surface smoothness of the target are both important factors in achieving efficient compression and heating [5]. However, in both indirect and direct drive, the design of the laser and the design of the fuel target cannot be decoupled. The target must be designed to avoid laser-plasma instabilities caused by, in particular, the Rayleigh–Taylor instability which occurs between the two fluids of different densities when they interact. In the case of the ICF fuel target, this happens when the outer layer of the target is ablated and changes state before the inner layer containing the D–T fuel. A key rationale of the indirect drive approach is that the hohlraum is designed to minimize such disruptions. However, in both approaches it is this challenge of laser-target interaction throws into question the viability of an ICF reactor altogether.

In addition to the design and manufacture of fuel targets (detailed in the next section on fuelling and exhaust systems), there are questions over whether it will be possible for an ICF fuel target to be struck with pinpoint accuracy by lasers with a very high frequency (perhaps ten laser pulses per second) [8]. The repetition rate of current laser technology must be improved dramatically for ICF approaches to be commercially viable [5]. Further, current neodymium lasers are typically less than 1% efficient at converting electrical energy into laser energy and, as such, there is also a question regarding the efficiency of ICF routes to fusion energy—akin to the challenges of using NBIs in MCF approaches. Novel lasers, such as lithium lasers, as well as alternative laser configurations (including Chirped Pulse Amplification (CPA), a technology being explored by a privately funded fusion company in

Germany called Marvel fusion, and a company in Australia called HB11 Energy (see chapter 7)), may be capable of delivering extremely high power over very short periods with higher efficiency. Further details on the progress of ICF approaches that utilize high-repetition laser systems are detailed in chapters 6 and 7.

By contrast, First Light Fusion's reactor concept (referred to earlier in this chapter) intends to harness the same instabilities that are being avoided in other ICF approaches. The company suggests this permits a simpler target design. First Light's concept uses an advanced implosion process in which the compression of a gas-filled target is initiated via a shockwave from a high-velocity gas gun to create the conditions required for fusion [14]. The approach still depends upon the development of a repeatable, high-power, and high-efficiency gas gun, which may be more viable than the development of improved lasers that are a necessity for other laser-based ICF approaches.

While the challenges for MCF and ICF plasma production and control are well understood, the problems for MIF are, by comparison, underexplored. Additionally, as illustrated by the range of possible MIF approaches in figure 2.4, the challenges involved with plasma production and control are wholly concept-specific. General Fusion's concept, for example, is to inject a plasmoid into a spherical reactor chamber, where it is promptly compressed and heated by a metal liner that is spherically imploded by an array of acoustically driven pistons. This is in contrast to the MagLIF approach, which uses a magnetic field to compress a cylindrical fuel target before it is then imploded using lasers [8, 18]. Whilst in the MCF space, for example, the development of high-performance HTS magnets could benefit all approaches, a technology breakthrough with one MIF reactor will not necessarily have an impact on an alternative MIF approach. Relatedly, however, it is worth pointing out that HTS magnets may be a key technology for some MIF approaches. However, HTS might be an important technology for some concepts, the mechanism for the driver, liner and fuel target for all MIF reactor concepts still require novel, concept-specific solutions. Similar to ICF, technology for MIF plasma production and control must be developed to be highly efficient and repeatable due to fact MIF reactors will be pulsed. Ergo, MIF reactors do have challenges that are similar to both MCF and ICF, with some potential for cross-cutting development, but the specific solutions required are quite different indeed.

2.3.2 Fuelling and exhaust systems

Fuelling systems are required to provide the deuterium and tritium fuel for the fusion reaction. After a fusion reaction has taken place—either in a single pulse in an ICF or MIF reactor, or in a continuous fashion in an MCF reactor—the products from the reaction (mainly helium), as well as unspent fuel and impurities, must be removed from the fusion reactor via an exhaust system. It is perhaps most instructive to think of the fuelling and exhaust systems on a fusion reactor like that of an internal combustion engine, in which fuel and air are injected, ignited, and compressed via a piston. The resulting combustion reaction drives the piston back to its start position to repeat, and the products of the reaction, as well as any unspent fuel, are exhausted, and the process is repeated cyclically[9].

In MCF reactors, fuel is injected via gas puffing or pellet injection. For a commercial fusion reactor, the fuel must be supplied at a rate sufficient to allow continuous, steady-state operation. Gas puffing introduces the fuel as a gas at the edge of the plasma. While there are few challenges associated with gas puffing technology, it has been widely tested in MCF experiments and is relatively well established. The issue with gas puffing is that as a technique it is limited as it cannot deliver fuel to the centre of a plasma. Pellet injection involves injecting solid pellets of frozen fuel at high velocity to thus allow insertion into the central region of the plasma. The angle and area in which the fuelling pellets are inserted, as well as the high efficiency of fuelling, are important challenges for pellet injection. However, these have been extensively explored for integration in the ITER tokamak and are at a relatively high level of technology readiness.

It is necessary to first understand how plasma behaves inside an MCF reactor before we can understand the needs of the exhaust system. The energy in the core of an MCF plasma transports outwards towards the edge of the plasma. The edge of the plasma can gather impurities from the reactor walls, contaminating the plasma, reducing its temperature, thus degrading overall plasma performance, and also leading to erosion of the reactor vessel [5]. MCF reactors thus typically include specific regions where the plasma can purposely make contact with a surface. This region is typically referred to as a divertor. It is named as such because its function is to deliberately 'divert' the edge of the plasma away from the plasma core and in the process transfer the eroded material to avoid diluting it. Most divertor designs allow the plasma to interact with a solid surface, although some liquid designs are now also being explored (as later detailed in chapter 10). A key function of the divertor is to remove the unspent fuel from the plasma and to separate it from helium 'ash', to then be transferred to the tritium systems where it can be processed and later reintroduced to the reactor as fuel[10]. A specific challenge is how efficiently the exhaust gases and particles can be separated.

As well as for plasma stability and for separating exhaust products, the divertor has a key function to handle the power load from the plasma, transferring it as heat which can be used for energy generation. All divertor surfaces are subjected to very high-power loads and high temperatures, which can incur heavy damage. Divertors must, therefore, be designed to have an acceptable operational lifetime that can withstand the high-power and particle loads. This is fundamentally a materials issue and is discussed in more detail in chapter 10 as part of the wider challenge to develop advanced fusion materials.

Fuelling and exhaust systems for ICF and MIF reactors share many similarities, particularly compared with MCF. Both involve the injection of a target which is

[9] This analogy works best for ICF and MIF reactors, as they are pulsed, while MCF reactors are steady-state and thus continuously fuelled.

[10] To process and recycle tritium and deuterium fuel is a primary function of the tritium systems, as later detailed in chapter 10.

then imploded. The fusion products, unspent fuel, and—specifically in the case of ICF—the debris from the imploded fuel targets are removed via an exhaust system, and the process is repeated cyclically. This is shown in the schematic in figure 2.5 which, although derived from an ICF configuration, broadly applies to MIF reactors too.

As noted in the earlier description of ICF plasma production and control, a key challenge is to inject the fuel target to be synchronized with the laser pulse [30]. The fuel target injector must be capable of maintaining synchronicity at a high repetition rate, as some ICF concepts have a design basis for ten shots per second [20, 28]. Additionally, fuelling systems must allow for in-flight tracking to measure position, flight direction, and velocity [29, 31, 32]. Synchronization of delivering the target and laser illumination has been demonstrated [33, 34], for which progress is detailed in chapter 6.

Then there is the design and manufacture of the fuel target itself. The design of a fuel target is dependent on implosion physics criteria as well as thermal, mechanical, and optical properties of the selected material [32]. However, since some ICF reactor designs may consume almost 1 million targets per day, the mass manufacture of fuel targets that are low cost and simple is an integral part of the overall reactor design. Indirect drive ICF experiments in the past have used gold hohlraums, due to the high x-ray conversion efficiency and ease of fabrication of using gold. The current cost, however, is US$ 2500 per target. For a commercially viable ICF reactor, fuel targets must be in the region of US$ 0.25 each [30]. This may be achievable using lead instead of gold, which is approximately 20 000 times cheaper, and which has been demonstrated to be capable of the same performance as gold [27].

As with MCF reactors, ICF exhaust systems must remove fusion products as well as unspent fuel (to be recycled), but it must also recover and process the debris of the imploded target from the reactor chamber [29, 32]. Unlike MCF, however, the exhaust system is not critical for plasma stability as in MCF reactors. Instead, ICF —as well as MIF—exhaust systems operate more like the exhaust in an ion cyclotron emission. Nonetheless, no ICF reactor has demonstrated a continuously operating target debris removal and fuel recovery system and it thus remains one significant technology development gap.

In contrast to typical laser-based ICF approaches, for which the above details are relevant, it is instructive to once more consider First Light Fusion as an example. As noted in the previous section, First Light is developing a fuel target with a simpler design that is low cost and easy to fabricate. For the exhaust system, First Light's reactor concept employs a liquid metal 'curtain' surrounding the reactor chamber that catches the debris from the imploded fuel target. The curtain also protects the reactor vessel by absorbing the energy from the reaction, which is carried by the fusion neutrons that are also used for breeding tritium (later detailed in chapter 10).

For MIF reactors, as in ICF reactors, it is the rate at which the fuel target or plasma can be injected that is critical, as they are pulsed systems. However, the required target injection frequency for a MIF reactor is not as high as for a ICF reactor. It is estimated that the target injectors, as well as the drivers for the liner systems, must fire one shot every 5 to 10 seconds—although, noting again the wide range of MIF concepts being developed, this is dependent on the specific reactor concept [20, 28]. Some MIF concepts are potentially advantageous over both MCF and ICF as regards fuelling. Solid fuel targets may not need to be manufactured—with MagLIF as an obvious exception—as in ICF. Similarly, a stable plasma with high-energy confinement time is not required as in MCF, as it is compressed and heated very soon after injection. Although magnets, required to compress the plasma target in some MIF concepts, are a clear cross-cutting technology between MCF and MIF that could rightly be considered part of the fuelling system, the issues for MCF reactors associated with steady-state operation, heating and current, plasma instability, and plasma impurity control (by exhausting), do not apply to MIF reactors. Again, these are some further reasons as to why MIF proponents believe that MIF reactors offer a potentially optimal route to fusion.

Whilst the fuelling systems for MIF reactors combine some aspects of MCF and ICF, the exhaust systems for a MIF reactor are—as shown by figure 2.5—largely similar to an ICF reactor. The MIF reactor chamber must be evacuated between shots at a rate consistent with the fusion pulses, i.e. once every 5–10 s. Unlike ICF reactors, however, using plasma targets means that many MIF concepts will not produce debris. The belief is, therefore, that MIF reactors will avoid the instabilities associated with MCF in continuously fuelling and exhausting a plasma, but will simultaneously avoid the solid target injection and debris removal associated with ICF reactor exhausts. Importantly, however, none of the proposed fuelling or exhaust designs for MIF reactors has been experimentally tested (to the best of the authors' knowledge). In contrast, the operational experience of MCF and ICF concepts has demonstrated at least fundamental technical viability.

2.3.3 Diagnostics

Measuring the value of the key parameters such as density, temperature, the level of impurity in the plasma, the magnetic field and plasma position (in the case of MCF) or target position (in the case of ICF and MIF) is fundamental to being able to operate and control a fusion reactor. A variety of diagnostics that can measure all types of radiation: electromagnetic (including emission, absorption, refraction, reflection and scattering), neutron, and energetic ion (particle) radiation are required for real-time feedback and control in a fusion reactor [5, 35]. Such diagnostics are also crucial for enhancing understanding of physics, by providing feedback from experiments to test and improve theoretical models. While an experimental reactor requires diagnostics to understand improve understanding of physics and reactor control, a commercial reactor will require diagnostics only to control the fusion reactor, not to fully understand the physics of the system (which by that point will be well understood). As such, a commercial fusion reactor will not require the same level of diagnostic equipment as, for example, ITER. A commercial reactor will, however, need additional diagnostics to monitor tritium breeding rate and temperature profile in the blanket, for example. These types of diagnostics challenges have received little attention but are similarly as essential as plasma diagnostics, yet they remain unresolved.

Fusion plasmas in reactors across all of the three conceptual spaces are different in scale and duration. This, in particular, has a significant effect on the type and characteristics of diagnostics systems required. Optical transmission materials, neutron detection systems, and electronics are required to operate in environments with high levels of neutrons, gamma-ray and x-ray radiation, regardless of the type of reactor. However, the specific diagnostics vary due to the differences in dose level, rate, energy spectra and durations of the radiation produced across the three conceptual spaces [35–37]. While some specific diagnostics systems cannot be shared across conceptual spaces, it is important here to first understand the conditions of the fusion nuclear environment before the specific challenges for reactors in each of the conceptual spaces can be detailed.

All fusion reactors will be subject to high temperatures and significant mechanical loads, including from the applied magnetic fields or the force of the fusion implosion, in the case of MCF and ICF or MIF, respectively. For instance, optical diagnostics closest to the fusion reaction will be subject to intense neutron and x-ray radiation, which will likely degrade the optical properties resulting in reduced component lifetime, or even failure, of those components [5, 38]. Optical systems—including fibres, mirrors, and windows—must, therefore, be developed to withstand the harsh fusion nuclear environment[11]. It is also important to understand that for a commercial reactor, in almost all cases, currently established diagnostics will be subject to more extreme conditions relative to the environments in which they have been qualified for today. The levels of radiation—in particular, the neutron heating, flux and fluence, particle bombardment, and pulse length—in ITER will be ten times higher than in JET, an experimental tokamak upon which aspects of ITER are based [35]. The transition to a next-step commercial demonstration reactor will see a larger jump, still. This may prevent the use of existing diagnostics and techniques, meaning new ones must be developed. The enhanced radiation effects will likely lead to much reduced lifetime of diagnostic equipment, along with—perhaps obviously—other components in the reactor vessel (see [35, 39]).

For MCF reactors, diagnostics are fundamental for measuring the properties of the steady-state plasma—the shape, position, current, density, impurity levels, required heating etc—to inform control and to keep it within defined and, particularly in the case of tokamaks, well understood limits [5, 35]. Diagnostics, therefore, have a function to control but also to detect, and subsequently to inform the control of the plasma to prevent disruptions that can critically damage the reactor. As with several of the other challenges, the diagnostics requirements for a commercial MCF reactor are well understood, principally due to the focus on diagnostics for ITER and satellite experimental reactors supporting the project[12]. Most broadly, improving diagnostics performance to withstand the more extreme

[11] Incidentally, the development of optical equipment is identified as a key cross-cutting diagnostics issue that affects all fusion reactor approaches (MCF, ICF and MIF), see [36, 43].

[12] Specifically, see [40].

environments (as indicated in the previous paragraph), while lowering the cost and reducing the size and number of diagnostics represent the key challenges in developing next-step MCF reactors towards commercialization [35, 40–42].

The diagnostics required for an ICF reactor are fundamentally different from those required for an MCF reactor. Principally, the accuracy of the laser striking the target (and thus, the position of the target) and the yield of the implosion require precise measurement [7]. Diagnostics to monitor and inform the control of the laser beams, in particular for positioning, as well as for the target manufacturing systems, will also be needed. As an example, the operation of NIF—as an experimental ICF reactor—employed 35 different diagnostics to measure a wide range of reactor parameters. The most important of these were for the characterization of x-rays striking the target, additional x-ray imaging to visualize the fuel target as it ignited, and gamma-ray detection and neutron spectroscopy to measure the fusion reaction yield upon implosion [7]. As with a commercial MCF reactor, a commercial ICF reactor will be subjected to higher levels of radiation than an experimental reactor. While an MCF reactor will be subject to greater levels of energetic ions and particles, however, diagnostics in an ICF reactor will be subject to an additional powerful electromagnetic pulse (EMP) produced during the inertial fusion reaction. EMPs can cause transient currents in electronic devices and are thus likely to affect the diagnostics surrounding the reactor chamber to give rise to spurious signals or damage. Additionally, debris from the imploded target, acting as high-velocity shrapnel, may damage optical surfaces inside the reactor chamber [36].

The requirements of a MIF reactor bear some similarity to both the MCF and ICF approaches. Typically, three diagnostic sets will be needed. For MIF concepts that require plasma formation and injection, the first set of diagnostics to support the formation of the target plasma is similar to those required for MCF. These diagnostics must measure the magnetic field, density, temperature, duration, and position of the plasma [43]. The diagnostics systems to create the plasma for the general fusion and the PLX concepts are similar to those required for a tokamak, for example (compare [20, 35, 44]). The second set of diagnostics is required to measure the implosion. These diagnostics must determine the position and speed of the driver as it compresses and heats the target plasma; for which the same parameters as before (field, density etc) must be continually monitored. In this second phase (liner compression by the driver), the plasma injector, driver and liner must each have its own set of diagnostics to allow for the control and synchronization of shockwave structure, thickness, and compression and heating rates [20]. The third and final phase—the 'burn' phase when the plasma is compressed, and fusion occurs, several diagnostics are required. Diagnostic instruments are required measure the position and pressure of the liner during and after the implosion, to measure the fusion yield (in a similar way to ICF), and also to measure the evacuation of the chamber after the pulse [20, 43]. As with ICF reactors, during the implosion of a MIF liner—where the target is rapidly compressed and heated—large electrical transients can be induced by the fusion EMP blast, which can affect the electronics in the diagnostics. The electromagnetic and mechanical forces from the target implosion during the second and third phases detailed here are particularly challenging [43]. Primarily, this is because all diagnostic challenges for MIF reactors are constrained by access to the reactor chamber, and, as with both MCF and ICF concepts, by the extreme temperatures and fusion nuclear environment [20]. Moreover, diagnostics to detect the position of the injected plasma are also particularly difficult to place physically inside the reactor vessel of some MIF concepts. Finally, as with the challenges in plasma production and control, and fuelling and exhausts, few MIF reactors have been developed to the point at which the level of diagnostics detailed herein can be tested. As such, the technology can be broadly considered to be at a low level of technical readiness and there may be additional diagnostics challenges that are not yet known.

2.4 Summary

This chapter has reviewed three conceptual spaces to fusion energy—magnetic confinement fusion (MCF), inertial confinement fusion (ICF), and magneto-inertial fusion (MIF)—and the differences between the reactors in each space. MCF reactors use sets of magnets to confine a fusion plasma, which is then heated to achieve fusion. ICF reactors achieve the conditions for fusion instantaneously by rapidly compressing and heating a small fuel target by using one or more high-power lasers. MIF reactors, to an extent, combine aspects of both MCF and ICF. MIF reactors create and inject a target, most commonly in the form of a plasma, into a reactor chamber where it is then compressed and heated rapidly by a converging liner (accelerated by a set of drivers) to achieve the temperatures and densities required for fusion. Reference is made to the most common configurations for MCF and ICF, including a description of the most well understood examples (the tokamak and the stellarator for MCF, and direct drive and indirect drive approaches for ICF). For MIF, reference is made to several approaches or concepts being pursued by private companies, as the development of fusion reactors in this space is largely the focus of the private sector, rather than publicly funded research.

Although the specific reactor configurations, and thus more broadly the paths to fusion energy, are varied, there are some fundamental commonalities in physics and engineering. The specific engineering challenges associated with core reactor systems—plasma production and control, fuelling and exhaust, and diagnostics—have been explored in-depth, with consideration of how the challenges differ for reactors in each of the conceptual spaces. Examples of developers pursuing MCF, ICF and MIF approaches, from both public and private sector programmes, are provided to underline the challenges in these areas. Overall, the chapter has provided a basis for the reader to understand the subsequent chapters of this book.

Supplementary Tables

Table 2.S1. List of MCF devices, assembled from [3].

Type	Ownership	Name of the device	Affiliation	Country	Status (as of September 2020)
Conventional tokamak	Public	NOVA-UNICAMP	UNICAMP Universidade Estadual de Campinas	Brazil	Operating
		TCABR	University of São Paulo	Brazil	Operating
		STOR-M	Plasma Physics Laboratory, University of Saskatchewan	Canada	Operating
		HT-7	Hefei Institutes of Physical Science	China	Operating
		J-TEXT	Huazhong University of Science and Technology	China	Operating
		EAST	Institute of Plasma Physics, Chinese Academy of Sciences	China	Operating
		HL-2A	Southwestern Institute of Physics	China	Operating
		COMPASS, COMPASS-U	Institute of Plasma Physics of Academy of Sciences	Czech Republic	Operating
		GOLEM	Institute of Plasma Physics of Academy of Sciences	Czech Republic	Operating
		EGYPTOR	Egyptian Atomic Energy Authority	Egypt	Operating
		ToriX	École Polytechnique	France	Operating
		WEST	IRFM-CEA	France	Operating
		ASDEX Upgrade	Max Planck Institute for Plasma Physics	Germany	Operating
		ADITYA-U	Institute for Plasma Research	India	Operating
		SST-1	Institute for Plasma Research	India	Operating
		Alvand	Iran Atomic Energy Organization	Iran	Operating
		Damavand	Iran Atomic Energy Organization	Iran	Operating
		IR-T1	Plasma Physics Research Center	Iran	Operating
		FTU	ENEA Frascati Research Center	Italy	Operating

HYBTOK-II	Nagoya University	Japan	Operating
TOKASTAR-2	Nagoya University	Japan	Operating
JT-60SA	National Institutes for Quantum and Radiological Science and Technology	Japan	Operating
PHiX	Tokyo Institute of Technology	Japan	Operating
LIBTOR	Tajoura Nuclear Research Center	Libya	Operating
ISTTOK	Instituto Superior Técnico	Portugal	Operating
KSTAR	National Fusion Research Institute	Rep. of Korea	Operating
FT-2	Ioffe Physical-Technical Institute of the Russian Academy of Sciences	Russia	Operating
TUMAN-3M	Ioffe Physical-Technical Institute of the Russian Academy of Sciences	Russia	Operating
T-11M	Troitsk Institute for Innovation and Fusion Research	Russia	Operating
TCV	Swiss Plasma Center	Switzerland	Operating
JET	Culham Science Centre (EUROfusion)	UK	Operating
HBT-EP	Columbia University Plasma Physics Laboratory	USA	Operating
DIII-D	General Atomics	USA	Operating
DTT	ENEA Frascati Research Center	Italy	Planned
PLATO	Kyushu University	Japan	Planned
HL-2M	Southwestern Institute of Physics	China	Under construction
ITER	ITER HQ	France	Under construction
T-15MD	Kurchatov Institute	Russia	Under construction
TT-1	Thailand Institute of Nuclear Technology	Thailand	Under construction

(*Continued*)

Table 2.S1. (*Continued*)

Type	Ownership	Name of the device	Affiliation	Country	Status (as of September 2020)
Spherical tokamak	Private	EXL-50	ENN	China	Operating
		ST40	Tokamak Energy Ltd	UK	Operating
		SPARC	Commonwealth Fusion Systems	USA	Planned
	Public	ETE	National Institute for Space Research-INPE	Brazil	Operating
		SUNIST	Department of Engineering Physics Tsinghua University	China	Operating
		MEDUSA-CR	Tecnológico de Costa Rica	Costa Rica	Operating
		QUEST	Advanced Fusion Research Center, Kyushu University	Japan	Operating
		TS-4U	Hongo (Yayoi) Campus, The University of Tokyo	Japan	Operating
		TS-6	Hongo (Yayoi) Campus, The University of Tokyo	Japan	Operating
		TST-2	Kashiwa Campus, The University of Tokyo	Japan	Operating
		UTST	Kashiwa Campus, The University of Tokyo	Japan	Operating
		LATE	Kyoto University	Japan	Operating
		KTM	National Nuclear Center of the Republic of Kazakhstan	Kazakhstan	Operating
		GLAST-III	Pakistan Atomic Energy Commission	Pakistan	Operating
		Globus-M2	Ioffe Physical-Technical Institute of the Russian Academy of Sciences	Russia	Operating
		GUTTA	Saint Petersburg State University	Russia	Operating

		LTX-β	Princeton Plasma Physics Laboratory	USA	Operating
		NSTX-U	Princeton Plasma Physics Laboratory	USA	Operating
		Pegasus-III	University of Wisconsin-Madison	USA	Operating
		SSST	Institute for Plasma Research	India	Planned
		MAST-U	Culham Centre for Fusion Energy	UK	Under construction
Heliotron	Public	Heliotron J	Kyoto University	Japan	Operating
		LHD	National Institute for Fusion Science	Japan	Operating
Stellarator	Private	Renaissance Fusion	Renaissance Fusion	France	Planned
	Public	SCR-1	Tecnológico de Costa Rica	Costa Rica	Operating
		Wendelstein 7-X	Max Planck Institute for Plasma Physics	Germany	Operating
		TJ-K	University of Stuttgart	Germany	Operating
		TJ-II	National Fusion Laboratory	Spain	Operating
		Uragan-2M	Institute of Plasma Physics National Science Center	Ukraine	Operating
		Uragan-3M	Institute of Plasma Physics National Science Center	Ukraine	Operating
		CFQS	Southwest Jiaotong University	China	Planned
		HSX	University of Wisconsin-Madison	USA	Planned
Spheromak	Private	IDCD (Imposed-Dynamo Current Drive)	CTFusion	USA	Operating
	Public	UH-CTI	Kyushu University	Japan	Operating
		UH-MCPG1	University of Hyogo, Himeji Campus for Engineering	Japan	Operating
		UH-MCPG2	University of Hyogo, Himeji Campus for Engineering	Japan	Under construction

(Continued)

Table 2.S1. (*Continued*)

Type	Ownership	Name of the device	Affiliation	Country	Status (as of September 2020)
Reversed field pinch	Public	KTX	University of Science and Technology of China	China	Operating
		RFX	Consorzio Rfx	Italy	Operating
		RELAX	Kyoto Institute of Technology	Japan	Operating
		EXTRAP T2R	KTH Royal Institute of Technology, Alfvén Laboratory	Sweden	Operating
		MST	University of Wisconsin-Madison - Department of Physics	USA	Operating
Pinch	Private	ZAP ENERGY	ZAP ENERGY INC.	USA	Operating
		MIFTI	Magneto-Inertial Fusion Technologies, Inc.	USA	Planned
	Public	Z Machine	Sandia National Laboratories	USA	Operating
		Zebra	University of Nevada, Department of Physics, Zebra Pulsed Power Lab	USA	Operating
Magnetic mirror machine	Public	GAMMA 10/PDX	Plasma Research Center, University of Tsukuba	Japan	Operating
		GDT	Budker Institute of Nuclear Physics	Russia	Operating
		GOL3	Budker Institute of Nuclear Physics	Russia	Operating
		CFR	Lockheed Martin's Advanced Development Program Facility	USA	Operating
		Pilot GAMMA PDX-SC	Plasma Research Center, University of Tsukuba	Japan	Under construction
Field reversed configuration	Private	FAT-CM	Nihon University	Japan	Operating
		C2-W (NORMAN)	TAE Technologies	USA	Operating
		The Fusion Engine	Helion Energy	USA	Planned
		PFRC	Princeton Fusion Systems and Princeton Satellite Systems	USA	Planned

Levitated dipole	Public	RT-1	Kashiwa Campus, The University of Tokyo	Japan	Operating
Simple magnetized torus	Public	TORPEX	Swiss Plasma Center	Switzerland	Operating
Dense plasma focus	Private	Focus Fusion	Lawrenceville Plasma Physics, Inc. DBA LPPFusion	USA	Operating

Table 2.S2. List of ICF devices, assembled from [3].

Type	Ownership	Name of the device	Affiliation	Country	Status (as of September 2020)
Laser fusion	Private	First Light	First Light Fusion Ltd	UK	Operating
		Innoven Energy	Innoven Energy LLC	USA	Planned
	Public	LMJ	CEA CESTA	France	Operating
		LFEX	Institute of Laser Engineering Osaka University	Japan	Operating
		GEKKO XII	Institute of Laser Engineering, Osaka University	Japan	Operating
		NIF	National Ignition Facility	USA	Operating
		OMEGA	University of Rochester Laboratory For Laser Energetics	USA	Operating
Inertial electrostatic fusion	Private	Horne Hybrid Reactor	Horne Technologies LLC	USA	Under construction
Pulsed power Generated electron beam	Private	IVR-3 (Thunderbird)	Proton Scientific, Inc.	USA	Operating

Table 2.S3. List of MTF devices, assembled from [3].

Type	Ownership	Name of the device	Affiliation	Country	Status (as of September 2020)
Magnetized target fusion	Private	Fusion Power Core	Compact Fusion Systems	USA	Planned
		PJMIF	HyperJet Fusion Corporation	USA	Planned
		General Fusion	General Fusion Inc.	Canada	Under construction
Beam-target	Private	AGNI	AGNI Energy Inc.	USA	Planned

References

[1] Pearson R J, Costley A E, Phaal R and Nuttall W J 2020 Technology Roadmapping for mission-led agile hardware development: a case study of a commercial fusion energy start-up *Technol. Forecast. Soc. Change* **158** 120064

[2] Rampal B, Chaisson J and Pearson R 2020 The rise of commercial fusion startups: a brief history of fusion *Clean Air Task Force* (https://catf.us/2020/06/the-rise-of-commercial-fusion-startups-a-brief-history-of-fusion/)

[3] International Atomic Energy Agency (IAEA) 2020 IAEA fusion device information system (https://nucleus.iaea.org/sites/fusionportal/Pages/FusDIS.aspx)

[4] Wurden G A *et al* 2016 Magneto-inertial fusion *J. Fusion Energy* **35** 69–77

[5] McCracken G and Stott P 2013 *Fusion: The Energy of the Universe.* (Waltham, MA: Academic)

[6] Guo H *et al* 2020 Innovative approaches towards an economic fusion reactor *Natl Sci. Rev.* **7** 245–7

[7] Moses E I 2009 Ignition on the national ignition facility: a path towards inertial fusion energy *Nucl. Fusion* **49** 104022

[8] Betti R and Hurricane O A 2016 Inertial-confinement fusion with lasers *Nat. Phys.* **12** 435–48

[9] Kitagawa Y *et al* 2015 Direct heating of a laser-imploded core by ultraintense laser-driven ions *Phys. Rev. Lett.* **114** 195002

[10] Mori Y *et al* 2016 Fast heating of imploded core with counterbeam configuration *Phys. Rev. Lett.* **117** 55001

[11] Sakata S *et al* 2018 Magnetized fast isochoric laser heating for efficient creation of ultra-high-energy-density states *Nat. Commun.* **9** 1–9

[12] Gong T *et al* 2019 Direct observation of imploded core heating via fast electrons with super-penetration scheme *Nat. Commun.* **10** 1–10

[13] Matsuo K *et al* 2020 Petapascal pressure driven by fast isochoric heating with a multipico-second intense laser pulse *Phys. Rev. Lett.* **124** 35001

[14] Ringrose T J *et al* 2017 A hypervelocity impact facility optimised for the dynamic study of high pressure shock compression *Procedia Eng.* **204** 344–51

[15] National Research Council 2013 *An Assessment of the Prospects for Inertial Fusion Energy.* (Washington, DC: National Academies Press)

[16] Kirkpatrick R C 2008 Magnetized target fusion—possible reactor systems *Current Trends in International Fusion Research: Proceedings of the Fifth Symposium* ed E Panarella and R Raman (Ottawa, ON: National Research Council of Canada) p 145

[17] Ryzhkov S V 2014 Current state, problems, and prospects of thermonuclear facilities based on the magneto-inertial confinement of hot plasma *Bull. Russ. Acad. Sci. Phys.* **78** 456–61

[18] Gomez M R *et al* 2014 Experimental demonstration of fusion-relevant conditions in magnetized liner inertial fusion *Phys. Rev. Lett.* **113** 155003

[19] Hsu S C *et al* 2012 Spherically imploding plasma liners as a standoff driver for magneto-inertial fusion *IEEE Trans. Plasma Sci.* **40** 1287–98

[20] Lynn A G, Merritt E, Gilmore M, Hsu S C, Witherspoon F D and Cassibry J T 2010 Diagnostics for the plasma liner experiment *Rev. Sci. Instrum.* **81** 10E115

[21] de Peretti M and Sabatier M 2008 Magnetized target fusion *Current Trends in International Fusion Research: Proceedings of the Fifth Symposium* ed E Panarella and R Raman (Ottawa, ON: National Research Council of Canada) p 141

[22] Chen F 2011 *An Indispensable Truth: How Fusion Power Can Save the Planet.* (New York: Springer)

[23] Costley A E 2019 Towards a compact spherical tokamak fusion pilot plant *Philos. Trans. R. Soc.* A **377** 20170439

[24] Clarke R H and Cai Z 2012 Helium and fusion energy *The Future of Helium as a Natural Resource* vol. 35 ed W J Nuttall, B A Glowacki and R H Clarke (Oxon: Routledge) pp 235–64

[25] Menard J E *et al* 2016 Fusion nuclear science facilities and pilot plants based on the spherical tokamak *Nucl. Fusion* **56** 106023

[26] McAdams R 2014 Beyond ITER: neutral beams for a demonstration fusion reactor (DEMO) *Rev. Sci. Instrum.* **85** 02B319

[27] Ross J S *et al* 2013 Lead (Pb) hohlraum: target for inertial fusion energy *Sci. Rep.* **3** 1–3

[28] Odette R and Zinkle S 2019 *Structural Alloys for Nuclear Energy Applications* (Amsterdam: Newnes)

[29] Nakai S and Mima K 2004 Laser driven inertial fusion energy: present and prospective *Reports Prog. Phys.* **67** 321

[30] Aleksandrova I V and Koresheva E R 2017 Review on high repetition rate and mass production of the cryogenic targets for laser IFE *High Power Laser Sci. Eng.* **5**

[31] Hockaday M *et al* 2011 *White Paper Compilation in Inertial Fusion Energy (IFE) Development (u)* (Los Alamos, NM: Los Alamos National Lab. (LANL))

[32] Reyes S *et al* 2013 Life tritium processing: a sustainable solution for closing the fusion fuel cycle *Fusion Sci. Technol.* **64** 187–93

[33] Komeda O *et al* 2013 First demonstration of laser engagement of 1-Hz-injected flying pellets and neutron generation *Sci. Rep.* **3** 2561

[34] Mori Y *et al* 2019 1-Hz bead-pellet injection system for fusion reaction engaged by a laser HAMA using ultra-intense counter beams *Fusion Sci. Technol.* **75** 36–48

[35] Costley A E 2010 Towards diagnostics for a fusion reactor *IEEE Trans. Plasma Sci.* **38** 2934–43

[36] Bourgade J L *et al* 2008 Diagnostic components in harsh radiation environments: possible overlap in R&D requirements of inertial confinement and magnetic fusion systems *Rev. Sci. Instrum.* **79** 10F304

[37] Wurden G A 2011 *Magneto-inertial Fusion (Magnetized Target Fusion)* (Los Alamos, NM: Los Alamos National Laboratory)

[38] Stork D and Zinkle S J 2017 Introduction to the special issue on the technical status of materials for a fusion reactor *Nucl. Fusion* **57** 92001
[39] Rubel M 2019 Fusion neutrons: tritium breeding and impact on wall materials and components of diagnostic systems *J. Fusion Energy* **38** 315–29
[40] Walsh M *et al* 2015 Integration of diagnostics on ITER *2015 IEEE 26th Symp. on Fusion Engineering (SOFE)* 1–8
[41] Walsh M *et al* 2011 ITER diagnostic challenges *2011 IEEE/NPSS 24th Symp. on Fusion Engineering* 1–8
[42] Federici G, Biel W, Gilbert M R, Kemp R, Taylor N and Wenninger R 2017 European DEMO design strategy and consequences for materials *Nucl. Fusion* **57** 92002
[43] Wurden G A *et al* 2001 Diagnostics for a magnetized target fusion experiment *Rev. Sci. Instrum.* **72** 552–5
[44] Delage M, Blondal D and Richardson D 2011 Developing acoustic magnetized target fusion: an overview of the science and development program at General Fusion *32nd Annual Canadian Nuclear Society Conf.*
[45] Kline J L *et al* 2019 Progress of indirect drive inertial confinement fusion in the United States *Nucl. Fusion* **59** 112018

IOP Publishing

Commercialising Fusion Energy
How small businesses are transforming big science
William J Nuttall, Satoshi Konishi, Shutaro Takeda and David Webbe-Wood

Chapter 3

Considerations for commercialization strategies for fusion energy

Shutaro Takeda and Sehila M Gonzalez de Vicente

3.1 Introduction

3.1.1 Background

Fusion energy has a game-changing potential in fulfilling part of the global demand for sustainable energy. The world needs low-carbon energy sources to combat climate change while meeting the rapidly growing global energy needs, as embodied by the Goal 7 of the Sustainable Development Goals (SDGs)—'Ensure access to affordable, reliable, sustainable and modern energy for all.' To meet this pressing demand, progressive fusion enterprises are seeking new pathways for the commercialization of this technology. Investors' expectations for such new pathways have beeb increasing over the last several years, and fusion enterprises have now attracted more than one billion USD of investment in total as of 2020 (see chapter 7).

Traditionally, public institutions with large-scale experimental devices[1] paved the way to the advancements of fusion science. However, the situation is changing: the private sector started to propose compact and innovative fusion devices based on decades of research efforts by the public sector. Formerly, small-scale fusion energy systems were deemed technically tricky due to the design constraints of conventional fusion approaches. However, due to the advances in both material science and theoretical treatises, more compact and progressive fusion reactor designs are being proposed. Forward-looking fusion start-ups are putting an effort into the development of fusion systems that have the potential to meet needs of the mid-21st Century. A new era for fusion energy, where faster commercialization would be

[1] The directory of the majority of fusion experimental facilities worldwide is available through the IAEA e-resource called Fusion Device Information System (FusDIS), which contains information on more than 100 fusion devices, both public and private, that are currently in operation, under construction, closed or being planned.

achieved through the public–private partnership, is now open and this path needs to be further explored.

This is one of the most exciting developments in the area of fusion research that brings new ideas of fusion energy to the field. The International Atomic Energy Agency (IAEA) has engaged with these innovative fusion start-ups through its technical workshops and publications.

3.1.2 Content of this chapter

Global cooperation is essential for the ultimate success of fusion energy. The IAEA has acted as a research facilitator by bringing scientists and experts together to exchange information and to identify critical issues. Furthermore, it has organized training programmes and coordinated research projects in the field. Currently, around 70 Member States are working together in fusion research under the IAEA aegis.

In response to the new developments in the private sector, the IAEA organized the First Workshop on Fusion Enterprises in June 2018 in Santa Fe, US[2]. The purpose of the workshop was to analyze the potential role of fusion energy in future markets and to understand how fusion enterprises can contribute to the commercialization of fusion as a reliable future source of energy. This workshop discussed new pathways towards the commercialization of alternative fusion energy systems, which offers a wide variety of insights to the theme of this book—commercializing fusion. This chapter reports the key takeaways of this workshop, outlining a roadmap for the development of fusion energy systems in the coming decades, bringing together expert opinions from the diverse range of disciplines that are essential to fusion's commercial success.

For more IAEA publications and materials relevant to the area of fusion research and technology, please visit https://nucleus.iaea.org/sites/fusionportal/

3.2 Future market

3.2.1 Energy demand forecasts

The next decades will likely see a dynamic increase in global energy demand, the global primary energy demand will increase by a quarter in 2040, according to the IEA (the International Energy Agency of the OECD) Stated Policies (STEP) Scenario [1]. Studies carried out by the US Energy Information Authority (EIA) predict the same trend. Non-OECD nations are forecasted to account for 69% of the 911 quadrillion British thermal units (Btu) global energy consumption by 2050 (Reference Case, figure 3.1), with Asia showing the most substantial growth in energy use among non-OECD countries [2]. Both the IEA's STEP Scenario as well as EIA's Reference Case predict a similar scenario, where global coal use remains constant, whereas the use of renewables, natural gas, and petroleum products increases monotonically [1, 2]. They project that almost two-thirds of global capacity will be from renewable energies in the same period, making it the predominant choice in the power sector.

[2] More information is available on: https://nucleus.iaea.org/sites/fusionportal/Pages/Enterprises/2018/BoA.aspx

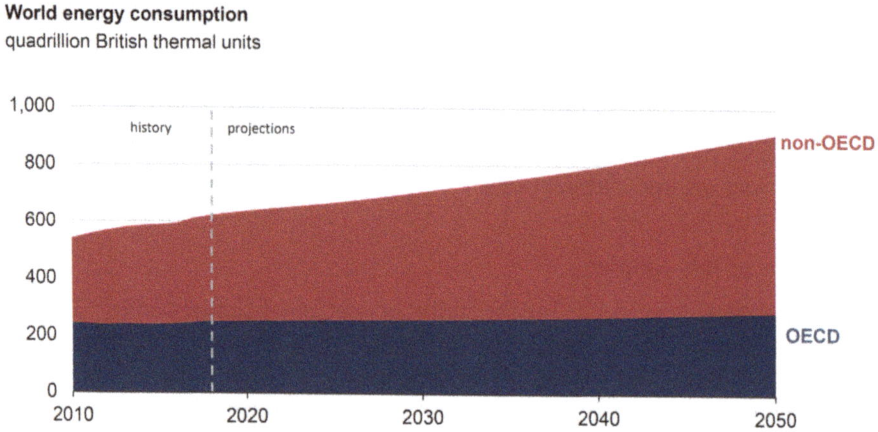

Figure 3.1. OECD and non-OECD energy consumption forecasts through 2050 (EIA Reference Case) [2].

3.2.2 The role of nuclear fusion in climate change mitigations

The global community is making efforts toward decarbonization to mitigate climate change under. To strengthen the global response, the Paris Agreement was adopted in 2015 with the main purpose of keeping the increase in the global average temperature well below 2 °C. Countries around the globe are planning their long-term strategies to achieve this goal. One such scenario, published by the European Commission, illustrates how challenging it is to achieve this goal. In this scenario, greenhouse gas emissions from electricity generation are reduced to zero by around 2040 (figure 3.2). These scenarios imply the transition to low-carbon sources may have to be completed by 2040 in the power sector.

Such a drastic reduction in carbon emission is a monumental task. Fusion energy, a sustainable energy source with low-carbon emissions, might be able to contribute to this goal if the technology were to be realized by 2030 as some start-ups claim. Early studies indicate that the life-cycle greenhouse gas emissions of electricity generation with fusion (~10 g CO_2-eq kWh^{-1}) is about half of that for fission (~20 g CO_2-eq kWh^{-1}) and about one-fifth of that of solar PV (~50 g CO_2-eq kWh^{-1}, depending on the location) [4, 5]. However, it has to be noted that a significant contribution is only possible by introducing a fleet of fusion energy systems, which might prove to be difficult by 2040.

3.2.3 Future market for fusion energy systems

Even though fusion energy has a powerful potential, future fusion power plants, when available on the market, will have to compete with other low-carbon power sources to be successfully commercialized. Potential competitors for fusion include renewable power sources combined with energy storage systems, advanced nuclear fission plants, and thermal power plants with Carbon Capture and Sequestration (CCS).

Finally, in addition to the electricity market, other potential markets for application of fusion systems may be attractive. These alternative markets include neutron sources for ordnance detection, space propulsion, waste transmutation, or

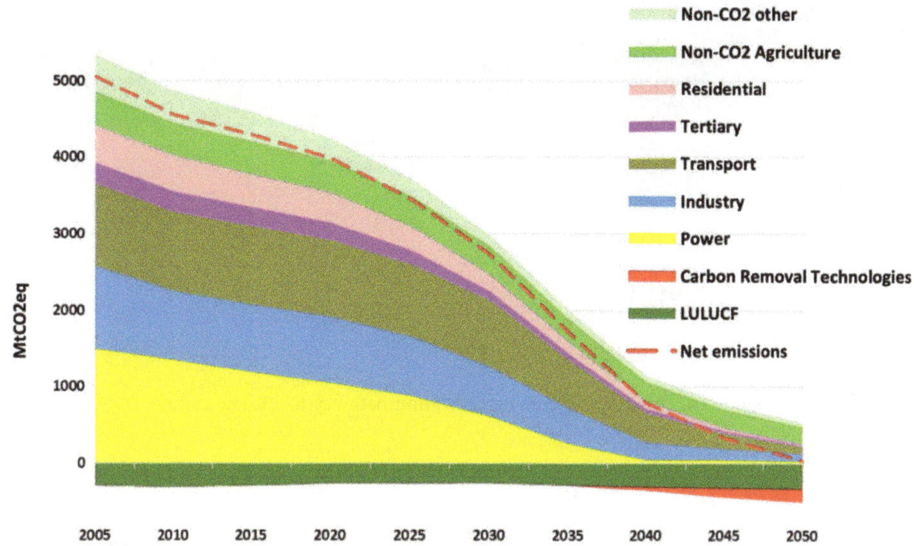

Figure 3.2. Greenhouse gas emissions trajectory by the European Commission for attaining the 1.5-degree target (LULUCF: Land Use, Land-Use Change and Forestry) [3].

Figure 3.3. 'Moonshot' commercialization pathway for fusion energy.

high-temperature heat application for fuel production, to name a few. It is important for the fusion community to consider these markets from the early stages of commercial fusion development.

3.3 Commercialization pathways

3.3.1 Four commercialization pathways

The term 'commercialization' can be defined as the process to introduce a new product to a market; in other words, it is the step from an R&D phase to a profit generating one. In the case of fusion energy, several different commercialization paths can be conceived. The four possible commercialization pathways for fusion energy, as proposed by Dr Eric Ingersoll, are illustrated in figures 3.3–3.6 and tables 3.1–3.4.

In the 'moonshot' commercialization scheme (figure 3.3), the technical demonstration of the concept is defined as the scientific breakeven, based on the previous development of the required technology to build a power-producing system. At this stage, funding usually comes from diverse sources like venture capitals, angel investors and the government (e.g., the US ARPA-E (Advanced Research Projects

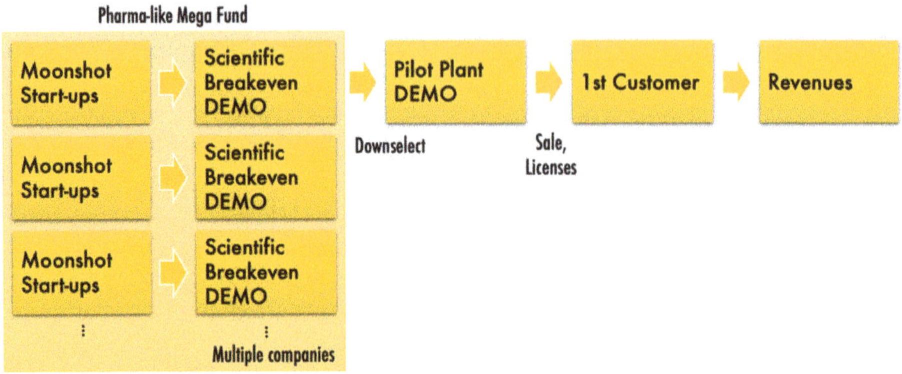

Figure 3.4. 'Mega-fund' commercialization pathway for fusion energy.

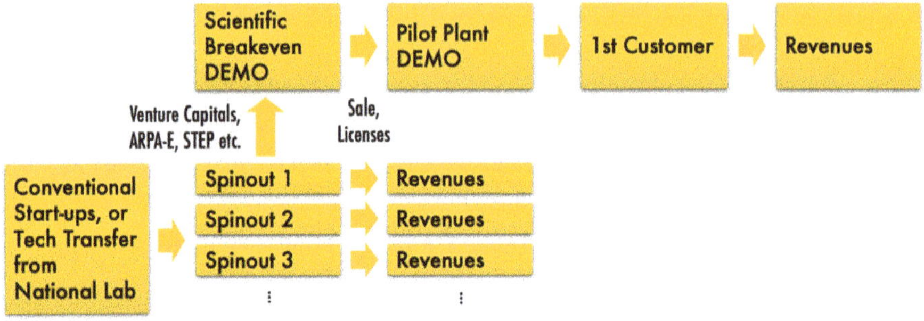

Figure 3.5. 'Spin-out' commercialization pathway for fusion energy.

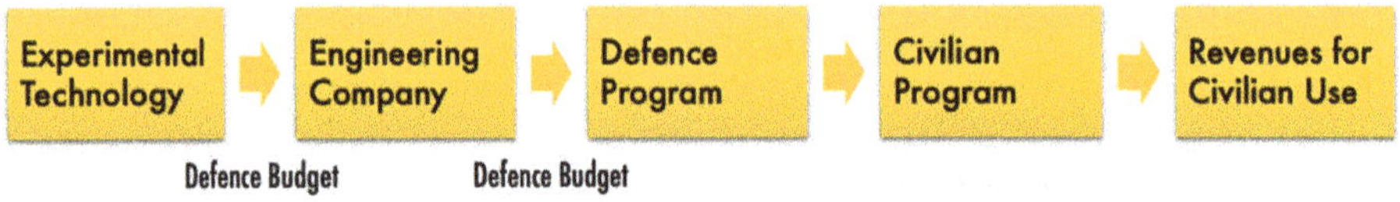

Figure 3.6. 'National emergency' commercialization pathway for fusion energy.

Table 3.1. Pros and cons of the moonshot pathway.

Pros	Cons
• Very focused approach; • Can cash-in early; • Shortest timeline to the goal.	• Needs firm support from Venture Capitals; • Revenues cannot be expected for more than 10 years; • Large initial capital risk.

Table 3.2. Pros and cons of the mega-fund pathway.

Pros	Cons
• Focused approach; • Synergy among concepts; • Large business ballast.	• IP issues may arise going from the first phase to the pilot phase; • Limited number of investors are capable of investing billions of USD; • Long time to revenues; • Many distinct concepts are needed.

Table 3.3. Pros and cons of the spin-out pathway.

Pros	Cons
• Can focus on necessary technology; • Fund raising from Venture Capital is easier with some technology; • Revenues targeted in shortest timeline.	• Some dilution of effort due to commercial focus; • Intellectual Property management could become complicated; • Longest timeline to fusion energy.

Table 3.4. Pros and cons of the national emergency pathway.

Pros	Cons
• Very focused approach; • Less stress for short-term investment.	• Defence, not civilian focused; • Revenues may never happen; • Longest timeline to revenues.

Agency–Energy) programme [6, 7], seeking to raise up to 100 million USD to demonstrate the technology). The movement to the next pilot plant stage requires a higher level of financial support, 500 million USD or more, that can come from the Public–Private Partnership. The drawbacks of this pathway are that any revenues are accrued only for the future and there is a lot of capital at risk.

The 'mega-funds' pathway (figure 3.4) has been typically pursued by the pharmaceutical industry, where investments are made in several companies of the same type until they reach the scientific breakeven. In the case of fusion, as of 2020, however, there is only one fusion-specific fund[3].

[3] Strong Atomics: https://strong-atomics.com/

In the 'spin-out' pathway (figure 3.5), risks are reduced by putting spinoff technologies into the market for other applications at an early stage. This pathway generates revenues sooner, reduces the required investments and enables the company to 'cash neutral' sooner. However, focus on the longer-term goals of achieving fusion energy may be blurred. So far, this model has proven successful to attract private funding. For instance, the Small Business Innovation Research (SBIR) programme [8, 9] has provided 150 thousand USD seed funds for a concept in Phase I[4] and up to 1.1 million USD for the development to commercialize the product in Phase II[5].

The government plays a significant role in the 'national emergency' pathway (figure 3.6) by linking the technological development to national security needs (mostly the defence). This model was followed by the US government to develop fission during World War II. Fusion energy systems are peaceful technology that has little/no link to military applications; however, fusion energy might follow this pathway under the new national emergency, that is climate change.

3.3.2 The benefits and the risks of spinoff businesses

The primary concern of investors in the fusion industry is that revenue from electricity generation with fusion energy systems may still be far away. For this very reason, some see spin-out businesses that generate nearer-term revenues as essential for step fusion enterprises. Cashflow is a critical matter, especially for start-ups, and earning near-term revenue from spinoff businesses would provide them with the much-needed cash for the continued research and development.

Possible spinoff business opportunities from fusion enterprises include high-temperature superconducting coils, fusion materials, neutron sources (for medical isotope production or radioactive waste transmutation) and fuel production (see chapter 12). However, fusion enterprises have to be aware that pursuing spinoff business would take up a considerable portion of the company resources, especially human resources, hindering the company from pursuing the primary mission of realizing fusion energy. Licensing related technologies (including the high-temperature superconducting materials) to other technological and industrial sectors may provide a good compromise, generating near-term revenues and gradual development of required infrastructure, while at the same time, not requiring too much company resources.

3.3.3 Technology readiness level

A product's technological readiness is often assessed with a scale named Technology Readiness Levels (TRLs), illustrated in figure 3.7 [10]. This scale was first introduced in 1974 by the National Aeronautics and Space Administration (NASA) to assess a new technology's maturity systematically:

- TRL of 1–2 indicates a basic technology research phase.
- TRL of 5–6 indicates a technology demonstration phase.
- TRL of 9 indicates the system test, launch and operations phase.

[4] Establishing the technical merit, feasibility, and commercial potential of the proposed technology.
[5] Continue the R/R&D efforts initiated in Phase I.

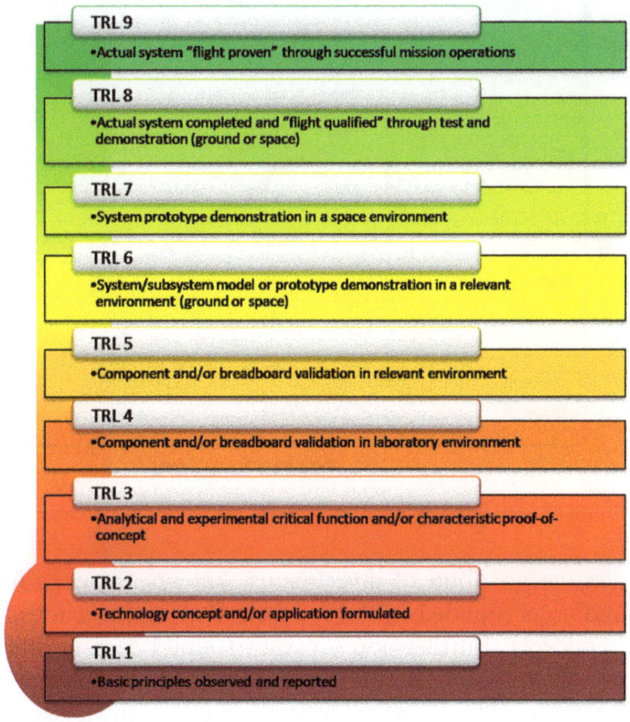

Figure 3.7. NASA Technology readiness level [12]. Courtesy of NASA.

ARPA-E supports innovative approaches in the TRL 1–2 phase, while private capitals are aiming to bridge the gap from TRL of 3 to 6. At the moment, many components of the fusion power plants are assessed as TRL 3 for conventional reactors [11].

3.3.4 Investment strategies for fusion enterprises

The commercialization of fusion energy is not likely to be achieved by simply following a single pathway. For instance, a company that is following the 'moon-shot' pathway can look for nearer-term revenues to satisfy investors by spinning out technologies.

Just as there is no single best pathway for commercialization, no single best strategy for investment exists for investors. Currently, investors are adopting a wide variety of investment strategies for fusion enterprises. Upon analysis, a commonality can be found in their strategies: they often reduce capital risks by a portfolio approach. In other terms, investors often invest in multiple promising concepts to increase the overall probability of success of their investments. This strategy also allows investors to gain access to a more significant number of markets while reducing the total capital expenditures.

Fusion investors may learn from successful examples in the pharmaceutical industry for their portfolio approaches. Investors in pharmaceutical companies often

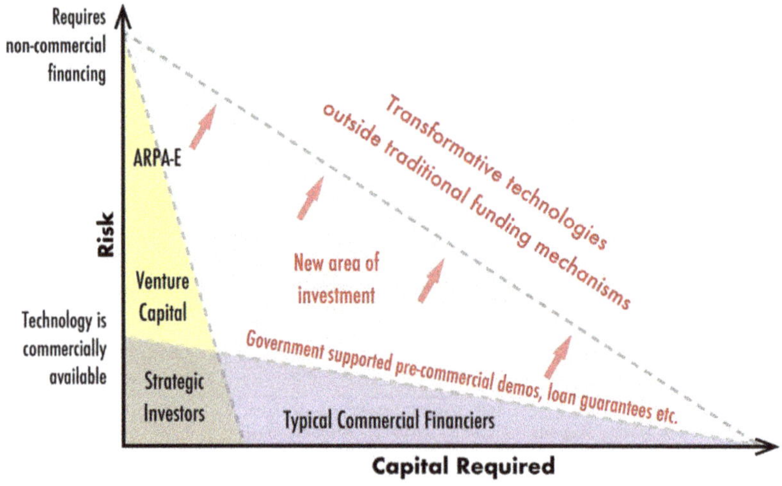

Figure 3.8. Types of investors plotted on risk versus capital required. Source: based on [13].

establish an extensive portfolio of investments, known as 'mega-funds.' The first fusion 'mega-fund' was founded in 2017, where the fund is currently investing in multiple fusion concepts. Furthermore, investors could seek another layer of risk reduction strategy by sharing essential equipment between concepts they invested, such as diagnostics or pulsed power systems. Similar strategies have also been adopted by public funding programmes in the last two decades, most notably by the US ARPA-E. Various types of investors in the market are illustrated by Dr Ingersoll on a plot of risk versus capital required in figure 3.8. This figure also outlines the area where new kinds of investments are needed to move transformative technologies to market.

Development of fission energy was primarily promoted as a matter of national security through the 'national emergency' pathway. As a result, commercialization of fission was achieved with little focus on economic viability, at least in its initial stages[6]. For fusion energy, the imminent need to combat climate change might give the fusion community a similar push for faster development.

3.4 Fusion power core design methodology

Over the past few decades, approaches for fusion reactor core designs have matured, and now there are some standard methodologies for the design of magnetic confinement fusion (MCF) systems. Fusion power cores are usually designed from the plasma outwards. First, the plasma operating modes are chosen based on performance and stability; then, the first wall is designed around the plasma; the blanket is built outside next; then the shield; and, finally, the magnetic systems. To satisfy all critical requirements, including the power deposition limits

[6] This is no longer the case, as the fission industry today is very focused on the cost to compete with renewable energy sources.

for the first wall and the Tritium Breeding Ratio (TBR) for the blanket, this process is repeated iteratively.

Conventionally, the majority of the fusion power cores for DEMO in the public programme are designed to produce 3–5 GW of fusion power, or 1–1.5 GW of electricity [14, 15]. However, in recent years, more and more alternative fusion power core designs with much smaller outputs, in the order of hundreds of MW of electricity, have been proposed by the private sector [16, 17]. These smaller fusion power cores are designed with economies of multiples in mind—the cost of such smaller fusion power cores may become competitive more quickly thanks to a deep learning curve and centralized manufacturing (i.e., they can be mass-produced as a module and shipped by rail or road).

The level of detail of such alternative fusion power core designs varies significantly concept by concept. Some are designed by a team of experienced scientists, while others are the brainchild of only a handful of visionaries. However, partly supported by the recent surge in financial investment, the maturity of such fusion power core designs has improved significantly in the last several years. Chapter 5–7 describe some of the notable alternative fusion power core designs; these chapters illustrate that engineers and scientists in the private sector are benefiting from design methodologies developed by the public sector, and they are now capable of designing fusion power cores for testing and beyond.

3.5 Constraints

When commercialized, many constraints will inevitably be imposed on fusion power plants, by regulatory licensing bodies, by the public, by policymakers and by utility companies. Therefore, extensive discussions have to be made on a wide range of fusion-specific concerns before commercialization, including the production and accounting of tritium, generation of radioactive waste, safety, electrical grid stability, security and legal aspects of fusion installations [18–21], to name a few. Fusion start-ups must be aware of these constraints from at the onset.

Among these constraints, discussion of licensing and regulatory frameworks is commencing. Some national regulatory authorities are already aware of fusion enterprises, some authorities have even engaged with private fusion start-ups. However, no collective strategies exist in the fusion community regarding how to engage with the regulatory authorities for the effective licensing frameworks for future fusion energy systems.

When should fusion enterprises engage with the licensing body? The experience from development of fission or other complex technologies suggests—the sooner the better, i.e. the private fusion start-ups need to start an informal dialogue with the licensing body as early as possible to discuss and agree on modalities of the engagement. Several enterprises have started this process already, including Shine and Phoenix in the US and General Fusion in Canada. In these cases, it was reported that processes have been generally smooth. Another notable example is ITER. The licensing process for ITER started much earlier than the completion of the engineering design [18]. This early engagement was necessary for timely approvals of preliminary to detailed engineering design stages, leading to the issue of a construction license. In the fission industry, a safety framework for Small Modular Reactors (SMRs) was successfully established in the US[7], which could also, after proper adaptation, be used by the fusion community as a model.

To reiterate, regulatory framework and licensing may be the primary unknown factor in the commercialization for many fusion enterprises both in terms of time and cost. There are at least two precedents for the community to learn from—ITER[8] and advanced fission reactors[9]. However, these two types of nuclear installations are not directly comparable, as ITER is an experimental plant, whereas the advanced fission reactors are already licensed commercial power plants. At the moment there is no consensus regarding how fusion start-ups should engage with regulatory authorities. Some experts argue that companies must start approaching authorities at the earliest stages (conceptual design) for successfully establishing effective frameworks.

3.6 Conclusions

Fusion energy could be a powerful tool to fight climate change as a virtually inexhaustible sustainable energy with low-carbon emissions if it could be commercialized by the mid-21st Century. Future tasks for fusion enterprises in striving toward fast commercialization of this promising energy, proposed by the IAEA Workshop participants, is summarized in the following.

(a) Engagement with national regulatory authorities to establish regulatory frameworks for fusion concepts at early stages:

mitigation of risk for investors is a key factor. In this sense, one of the primary unknown factors for fusion commercialization is the regulatory framework and licensing. The fusion community needs to engage in discussion with regulatory authorities authorities globally, preferably in a harmonized manner, as early as possible to establish a baseline for licensing discussions and to define as much as possible a licensing paradigm. It has to take into account the different designs for fusion systems, understanding common features and dissimilarities among them, and their potential impact in the regulation to be established. However, it is important to note that the licensing of a specific nuclear installation remains a national endeavour.

(b) Reinforcement of international coordination for safety, security and radioactive waste management for fusion power systems:

the establishment of the framework for safety, security and radioactive waste management should follow the same path as fusion development has followed from the very beginning—through international coordination. International coordination would greatly benefit the development of fusion power plants and its supply chains.

[7] 10 CFR Part 53: Risk-Informed, Technology Inclusive Regulatory Framework for Advanced Reactors Rulemaking

[8] Most of the work was carried out in the French context with the involvement of ASN and IRSN.

[9] Most of the work was done in the US by NRC and in Canada by CSNC.

(c) Actively inform the public about the potentialities of fusion energy as a tool to fight against global climate change, including how and when fusion energy could start making contributions:

This action would improve the public acceptance of this energy source and attract more investors. Investors are prone to fund more sustainable assets, so-called 'sustainable investment' or ESG investment (Environmental, Social and Governance). Corporate investments in new energy technology companies are growing strongly, reaching their highest ever level of over 6 billion USD in 2017. Fusion should be seen by investors as one of those sustainable assets with a high potential, in order to attract investment funds, including those from oil, gas and electricity utility companies. More work is needed by the community to differentiate fusion energy from other low-carbon energy options.

(d) Public–private partnerships should become a common model of the joint venture for the development of fusion energy systems:

the creation of joint ventures involving both private and public institutions in the fusion community would generate synergies that certainly would add value in many processes and benefit both parties. The development of fusion energy systems refers not only to the production of fusion power core concepts but also to a large number of enabling technologies (high-temperature superconductors, to take an example), where small enterprises can significantly contribute.

(e) Fusion enterprises should expand their horizon to international markets, both in developed and developing nations, including non-electricity markets:

As is shown in the EIA Reference Scenario, non-OECD countries will make up almost 70% of the total electricity consumption in 2040, implying a large market to be explored by fusion enterprises. Also, there are business opportunities for fusion enterprises in non-electricity markets, like advanced neutron sources that should be explored as a way to diversify revenues in a shorter term, therefore attracting investors.

Acknowledgment

The authors gratefully acknowledge the session chairs and participants of the First IAEA Workshop in Fusion Enterprises. In particular, the authors thank Dr Eric Ingersoll (contributed to the Commercialization Pathways section), Dr Ryan Umstattd (contributed to the Fusion Power Core Design section) and Dr Simon Woodruff (contributed to the Constraints section), and Dr Ronald L Miller (contributed to the Conclusions section).

References

[1] International Energy Agency 2019 *World Energy Outlook 2019* (Paris: IEA) (https://iea.org/reports/world-energy-outlook-2019).

[2] U.S. Energy Information Administration 2019 *International Energy Outlook 2019* (Washington, DC) (https://eia.gov/outlooks/ieo/pdf/ieo2019.pdf)

[3] European Commission 2018 *A Clean Planet For All: A European Strategic Long-term Vision for a Prosperous, Modern, Competitive and Climate Neutral Economy* (Brussels: EC) (https://eur-lex.europa.eu/legal-content/EN/TXT/?uri=CELEX%3A52018DC0773)

[4] Uemura S, Yamazaki K, Arimoto H, Oishi T and Shoji T 2009 Environmental assessment of fusion reactors by the calculation of CO_2 emission amount *Trans. At. Energy Soc. Japan.* **8** 34–9

[5] Tokimatsu K, Hondo H, Ogawa Y, Okano K, Yamaji K and Katsurai M 2000 Evaluation of CO_2 emissions in the life cycle of tokamak fusion power reactors *Nucl. Fusion* **40** 653

[6] Bonvillian W B and Van Atta R 2011 ARPA-E and DARPA: applying the DARPA model to energy innovation *J. Technol. Transfer.* **36** 469

[7] Nehl C, Umstattd R, Regan W, Hsu S C and McGrath P 2019 Retrospective of the ARPA-E ALPHA fusion program *J. Fusion Energy* 1–16

[8] Lerner J 1996 The government as venture capitalist: the long-run effects of the SBIR program *National Bureau of Economic Research* Report No. 0898–2937

[9] Audretsch D B, Link A N and Scott J T 2019 Public/private technology partnerships: evaluating SBIR-supported research *The Social Value of New Technology.* (Cheltenham: Edward Elgar Publishing)

[10] Mankins J C 1995 Technology readiness levels *White Paper* **6** 1995

[11] Tillack M, Turnbull A, Waganer L, Malang S, Steiner D and Sharpe J *et al* 2009 An evaluation of fusion energy R&D gaps using technology readiness levels *Fusion Sci. Technol.* **56** 949–56

[12] National Aeronautics and Space Administration *Technology Readiness Level 2012* (https://nasa.gov/directorates/heo/scan/engineering/technology/txt_accordion1.html)

[13] Ingersoll E (ed) 2018 A portfolio approach to funding and commercialization *1st IAEA Workshop on Fusion Enterprises*

[14] Federici G, Bachmann C, Biel W, Boccaccini L, Cismondi F and Ciattaglia S *et al* 2016 Overview of the design approach and prioritization of R&D activities towards an EU DEMO *Fusion Eng. Des.* **109** 1464–74

[15] Tobita K, Nishio S, Enoeda M, Nakamura H, Hayashi T and Asakura N *et al* 2010 *Conceptual Design of the SlimCS Fusion DEMO Reactor.* (Naka: Japan Atomic Energy Agency)

[16] Marmar E, Brunner D, Greenwald M, Hartwig Z, Hubbard A and Irby J *et al* 2018 The high field tokamak path to fusion energy: C-Mod to SPARC to ARC *60th Annual Meeting of the APS Division of Plasma Physics*

[17] McNamara S and Team T E 2019 Tokamak energy and the high-field spherical tokamak route to fusion power *APS Division of Plasma Physics Meeting 2019.* 2019: BO8. 014

[18] Perrault D, Niel J-C and Bruna G 2017 Nuclear fusion reactors *Safety and Radiation Protection Considerations for Demonstration Reactors that Follow the ITER Facility.* (Paris: IRSN) (https://irsn.fr/EN/Research/publications-documentation/Scientific-books/Documents/ITER-VA_web_non_imprimable.pdf)

[19] Takeda S, Konishi S, Yamamoto Y, Kasada R and Sakurai S 2017 Dynamic simulation-based case study of fusion on small-scale electrical grids *Fusion Sci. Technol.* **68** 341–5

[20] Takeda S, Sakurai S, Yamamoto Y, Kasada R and Konishi S 2016 Limitation of fusion power plant installation on future power grids under the effect of renewable and nuclear power sources *Fusion Eng. Des.* **109** 1754–8

[21] Pearson R J, Costley A E, Phaal R and Nuttall W J 2020 Technology roadmapping for mission-led agile hardware development: a case study of a commercial fusion energy start-up *Technol. Forecast. Soc. Change* **158** 120064

Chapter 4

Funding and financing commercial fusion power plants

David Webbe-Wood

4.1 Introduction

Whilst the question of whether small-scale fusion reactors can be made to work from an engineering and technological standpoint is considered in other chapters, this chapter considers the question as to whether such an endeavour would make sense from a financial or economic point of view. Any approach to commercial fusion will require substantial investment. This investment will not be forthcoming unless those providing the investment (whether governments, companies, or others) believe that a benefit is likely to be received in return. This benefit need not be financial, but may be another form of societal good. For all investors, an early return is preferable to one received later. However, as described by Offer [1], different parties will have different time horizons and the requirement of a commercial organization for a quick return means that some form of support from the Government may be required.

The current 'mainstream' international approach to commercial fusion is large scale and involves multiple partners. It is possible that small-scale approaches will be more amenable to attracting the necessary investments. However, proponents of commercial projects believe that commercialization can be achieved with a 'smaller, quicker, cheaper' approach [2]. It is possible that this approach will be more attractive to investors.

In comparison with some other methods of energy generation, nuclear energy, whether fission or fusion, suffers from the problem that major expenditure is required during the construction phase and then less expenditure is required during the operational phase when income from the sale of the energy (via whatever means) is received. This means that the cost of capital is a significant contribution to the overall costs. Compact reactors have been proposed as one means of reducing the costs of capital in the case of fission reactors and is an approach that has attracted subsidy

doi:10.1088/978-0-7503-2719-0ch4

from Governments (e.g. the United Kingdom Advanced Nuclear Technologies programme [3])

Proponents of compact reactors describe the advantages as:

> Small modular reactors offer a lower initial capital investment, greater scalability, and siting flexibility for locations unable to accommodate more traditional larger reactors. They also have the potential for enhanced safety and security compared to earlier designs [4].

In particular, the use of a modular approach enables much of the fabrication to be carried out in factory conditions, with the consequent increase in efficiency. The deployment of a fleet of power plants of the same design will also lead to savings as a result of learning effects.

4.2 Possible sources of funding

Even a small-scale fusion power plant will require a significant capital investment. Bechtel and collaborators [5] have estimated the overnight capital cost[1] of a 10th of a kind fusion generation plant with an output of 150 MW(e) to be in the range of $0.7–$1.9 billion[2]. These costs (and the costs of capital) will have to be provided prior to any income being received. There are a number of possible sources for this investment. Whichever source provides the capital investment, it will need to be confident that there is a reasonable likelihood of a return on its investment. This return may be either financial or in the case of Governments, a wider societal benefit (e.g. reductions in CO_2 emissions, increased energy security or providing support to a technology with potential for export earnings).

4.2.1 Provision of funds by a commercial operator

A large utility company might consider a compact fusion reactor as a part of its portfolio of energy generating assets. Any proposal to invest funds will have to compete with other proposed uses of the utility company's funds. These alternative proposals may well include compact fission reactors which share many (but not all) of the advantages of compact fusion reactors. As suggested by Zablielski [6] the attractions of using proven technologies may mitigate against the selection of fusion. This investment by a commercial company could be carried out through the use of a special purpose vehicle. As described by Sainati and colleagues [7] a special purpose vehicle is a fenced entity which enables the isolation of the assets, liabilities and risks within the vehicle. The vehicle also provides the project with remoteness from the risk of bankruptcy of the parent organization. The SPV is financed without (or with very limited) guarantees from the sponsors, such that lenders to the SPV depend on future project cash flows only and cannot recourse on the sponsor's other businesses [8].

[1] That is, the costs excluding the costs of capital.

[2] 2016 dollars.

4.2.2 Funds provided by a bank or similar investor

Landberg and colleagues [9] report that the global total of assets under management was $92 trillion in 2018 and that at least $30.7 trillion of this was held in green or sustainable investments, an increase of 34% since 2016. Whilst small-scale fusion may appear to be a suitable recipient of such investments there is, as Landberg *et al* point out, still no consensus as to what constitutes a 'green' or 'sustainable' investment

The European Union has developed an action plan on sustainable finance. As part of this action plan, an expert group has been established to develop a taxonomy of appropriate technologies. The EU Expert Group has excluded nuclear energy from their taxonomy of technologies [10] that aim to provide performance criteria for those bodies wishing to access 'green funding'. This exclusion is justified on the basis of the lack of final disposal routes for spent fuel. The Expert Group state that fusion (together with other technologies that are currently at low Technology Readiness Levels) may be added to the taxonomy in the future. Whether this updating of the taxonomy will occur and whether the timescale will be such that it benefits the development of small-scale fusion remains to be seen.

The EU also hopes that their taxonomy will impact on investors outside the EU. This absence of fusion from the taxonomy may make fusion ineligible to access 'green funding' more widely.

Banks, including 'green banks' need to balance the risk of the entity to which they are lending not being able to repay the loan against the interest rate charged. As stated above small-scale fusion is an untested technology and therefore will attract high interest rates. This coupled with the cost profile (high costs initially) and the extended times (circa 60 years) over which income to repay the funding will be received, means that the cost of capital will be a high proportion of the total costs.

The interaction between the level of risk and the rate of return that an investor would require to offset this risk coupled with the cost profile of a compact fusion reactor mean that the proposition may not be attractive to commercial investors. In this case funding may need to be provided by other sources such as Governments[3] directly or indirectly.

4.2.3 Funding provided by government

In the current economic circumstances (summer 2020) Governments are able to borrow at very low rates of interest (in some cases negative rates)[4] [11]. This means that the cost of capital for a project funded by the Government would be significantly lower than that of one reliant on commercial investment[5].

To some extent, Government funding has already occurred. Tokamak Energy have received funding from the UK Government under the Advanced Modular

[3] Which may include several Governments in a supranational organization.

[4] This might change following the aid programmes announced following the covid-19 outbreak

[5] Funding from current expenditure rather than from borrowing would effectively mean that the cost of capital was zero.

Reactor programme [3]. Similarly, General Fusion has received funding from the Canadian Government [12].

If a Government were to procure a compact fusion reactor power plant, it would have options as to how the plant would be operated. Prior to the partial privatization in 1990 all the UK's fission power stations were built for and operated by the Government owned Central Electricity Generation Board or the South of Scotland Electricity Board. A similar model could be applied to a fusion fleet. Alternatively, as suggested by the International Project Finance Association [13] for fission new build, a Government could procure the power plant, paying as costs arise and then sell the plant to the private sector once operational. The advantage of this approach, to the private sector, is that the Government has taken on the risk of the costs escalating or the performance not being what was expected. Therefore, the commercial operator will be able to access capital without having to pay a risk premium.

A variation to this approach would be for the Government to retain ownership of the power plant and to invite bids from private sector organizations to operate the plant for a period of time in a 'Government Owned Contractor Operated' arrangement in return for the income from the sale of the energy. Again, the Government taking on the risks associated with construction will lower the cost of capital.

Alternatively, a Government could provide support to private sector developers through providing capital at a rate lower than that at which the private sector developer would be able to obtain from commercial sources. The Government would take on the risk of the loan not being repaid. This support may be regarded as State Aid. State Aid is generally[6] not permitted in the European Union. World Trade Organization rules, however, do not generally prohibit such aid. [14]. Which rules the United Kingdom will be subject to, once exit from the European Union is completed at the end of the transition period remains to be seen.

A Government could also take on the risk by providing a guarantee for loans taken on by a private developer, agreeing to repay the loan if the developer is unable to do so. Thus, the provider of the loan would be relieved of the risk of default and would not charge the developer a risk premium.

Several politicians and political parties, as a response to increasing awareness amongst the electorate of climate change issues, have advocated 'Green New Deals'. Some of these proposals e.g. that of Joe Biden [15] include support for nuclear power. Thus, it is possible that Governments may be more inclined to fund fusion projects in the future.

4.2.4 Funds provided by the consumer

An alternative approach would be to transfer the risk or risk premium to the end consumer. This approach has been used or considered for use in the context of fission power plants.

[6] Exemptions to this prohibition are allowed, inter-alia, to support policy aims such as environmental protection and to support research, development, and innovation.

A contract for a different approach has been used for the Hinkley Point C power plant in the UK. In this approach the developer has been guaranteed a 'strike price'[7] for the electricity generated for 35 years. If the wholesale price is less than this strike price the developer will receive top-up payments which are ultimately recovered from the end consumer of the electricity. In the unlikely event that the wholesale market price of the electricity is greater than the strike price, the developer will pay the difference to the counterparty[8] [16]. Thus, the risk of low wholesale prices has been transferred from the developer. However, under this model the risk associated with cost escalations remains with the developer.

Recently, in the context of building fission power plants, the UK Government has consulted on and is currently (April 2020) considering the responses to, proposals for the adoption of a Regulated Asset Base (RAB) model [17] an approach which has been used for other infrastructure projects. The use of the RAB approach for nuclear having previously been proposed by Dieter Helm [18]. Under this model the Government will provide guarantees to the operator against specified low probability but high impact events, cost overruns would only be covered by this guarantee above a remote threshold. The operator would be allowed (by a Regulator) to charge Allowed Revenue payments to the customer to cover allowed costs, including the cost of capital, during both construction and operation. This transfer of risk to the Government and consumer means that the cost of capital will be reduced, receiving payments from consumers during the building phase will reduce the scale of the finance required.

Newbery and colleagues [19] have proposed a hybrid RAB model, similar to that used for the Thames Tideway Project, for fission nuclear generation projects in the UK. In this model, there is sharing of cost overruns between the developer and the Government/consumers. This, the authors believe, will encourage the participation of infrastructure investment funds. This coupled with a 'book building' exercise will further reduce the cost of capital.

When the consumers of the energy are industrial consumers or utilities financial models such as the Finnish Mankala approach may be appropriate. In this approach users of the output form a company to fund the construction and receive the output at cost price in proportion to their investment [20]. Lindroos and colleagues [21] report that Teollisuuden Voima Ltd which owns two fission power plants and is commissioning a third, obtained capital at an interest rate of less than 2.5% in 2016/2017 through this model.

As well as the RAB model, Dieter Helm has suggested a capacity market model [22] in which a nuclear plant developer could bid in an auction to provide capacity, as opposed to power, in a future period prior to the start of construction. If the developer were successful in the auction it would have a guarantee of income once constructed (the developer would, however, be left with the risk of having to provide the capacity from other sources if the development was unsuccessful or delayed).

[7] Which will increase in-line with inflation.

[8] In this case the counterparty is the Low Carbon Contracts Company which is a company wholly owned by the Secretary of State for Business, Energy and Industrial Strategy.

In whatever way a compact fusion reactor is funded those providing the funding will require a return (or at least a reasonable prospect of a return) on their investment. This return will be financial in the case of commercial investors. Governments may be willing to receive their return, at least in part, in the form of wider benefits such as increased energy security or reduction of the emissions of greenhouse gasses.

4.3 Sources of income

4.3.1 Income from sale of electricity

The UK is committed to net zero CO_2 emissions by 2050 [23]. This will mean the use of fossil fuels for the generation of electricity (at least when not coupled with carbon capture and storage (CCS) facilities) will no longer take place and low carbon methods of generation will be required. Some of this need will be met by renewable technologies such as wind and solar. Wind and solar energy suffer from the issue of interruptability. Currently this can be compensated for by the use of flexible gas or other fossil fuel generation, this will not be feasible in a net zero world. In the UK the Committee on Climate Change [24] has, in its further ambition scenario, estimated that there will be a need for 38% of the nation's electricity to be generated by 'firm' (i.e. always available) low carbon means by 2050.

The CCC also project that the UK's demand for electricity will approximately double (from 300 TWh in 2007 to 594 TWh) if the goal of net zero is to be met by 2050.

4.3.2 Income from other uses of the energy

As described elsewhere in this book, small-scale fusion is still a number of years away. Other low carbon techniques (including nuclear fission) are available, these technologies may mean that there will be no opportunity for small-scale fusion to contribute to the decarbonization of electricity generation. However, proponents of small-scale fusion believe that there will be a contribution to be made.

However, electricity generation is only part of a modern economy's energy needs. In the UK electricity was 17% of energy use in 2018 [25]. Other uses of the energy produced from a small-scale fusion reactor are described elsewhere in this book. For some of these applications the nature of fusion power plants gives them advantages when compared to alternatives. These may provide more lucrative opportunities for small-scale fusion operators.

4.4 Conclusions

The novel nature of fusion, being an as yet unproven technology, combined with the cost and income profiles mean that traditional methods of project finance may not be available to potential operators of compact fusion reactors. However, as outlined above a range of alternative funding models are potentially available. Some of these, however, will require the involvement of Governments for them to be available.

Acknowledgment

David Webbe-Wood wishes to thank the UK's Engineering and Physical Sciences Research Council for support through the ICO Centre for Doctoral Training in Nuclear Energy (Grant : EP/L015900/1).

References

[1] Offer A 2018 Patient and impatient capital: time horizons as market boundaries *Discussion Papers in Economic and Social History Number 165* August 2018

[2] Gryaznevich M P, Chuyanov V A, Kingham D and Sykes A 2015 Advancing fusion by innovations: smaller, quicker, cheaper *J. Phys. Conf. Ser.* **591** 012005

[3] BEIS 2018 Advanced Nuclear Technologies (https://gov.uk/government/publications/advanced-nuclear-technologies/advanced-nuclear-technologies)

[4] Office of Nuclear Energy 2016 Benefits of Small Modular Reactors (SMRs) | Department of Energy [Internet] pp 27–29 (https://energy.gov/ne/benefits-small-modular-reactors-smrs)

[5] Bechtel National Inc. WSI and DS 2017 Conceptual cost study for a fusion power plant based on four technologies from the DOE ARPA-E ALPHA Program (http://woodruffscientific.com/pdf/ARPAE_Costing_Report_2017.pdf)

[6] Zabielski V Regualted asset base – a model for the nuclear future *Power Eng. Int.* 24 April 2019. https://www.powerengineeringint.com/nuclear/strategic-development-nuclear/feature-regulated-asset-base-a-model-for-the-nuclear-future/

[7] Sainati T, Brookes N and Locatelli G 2017 Special purpose entities in megaprojects: empty boxes or real companies? *Proj Manag J.* **48** 55–73

[8] Steffen B 2018 The importance of project finance for renewable energy projects *Energy Econ.* **69** 280–94

[9] Landberg R, Massa A and Pogkas D 2019 Green finance: what it is and why it's growing (https://bloomberg.com/graphics/2019-green-finance/)

[10] EU Technical Expert Group on Sustainable Finance 2020 Financing a sustainable European economy *Taxonomy Report:Technical Annex* (https://ec.europa.eu/info/sites/info/files/business_economy_euro/banking_and_finance/documents/200309-sustainable-finance-teg-final-report-taxonomy-annexes_en.pdf)

[11] The Financial Times 2020 Latest bond rates, interest rates, libor and interbank rates – FT.com *Financial Times* (http://markets.ft.com/Research/Markets/Bonds)

[12] GeekWIre Canada invests $38M in general fusion to boost energy research (https://geekwire.com/2018/canadian-government-invests-38m-general-fusion-boost-energy-research)

[13] IFPA 2019 Response by IPFA membership to the Department for business, energy and industrial strategy on the RAB model for nuclear consultation. Vol. 3592310. (https://ipfa.org/wp-content/uploads/2019/10/Response-by-IPFA-membership-to-the-BEIS-consultation-on-the-RAB-model-for-new-nuclear_14.10.19.pdf)

[14] Jozepa I 2019 EU state aid rules and WTO subsidies agreement (https://commonslibrary.parliament.uk/research-briefings/sn06775/)

[15] Biden J 2020 Plan for climate change and environmental justice (https://joebiden.com/climate/)

[16] National Audit Office 2017 Report by the comptroller and auditor general – Hinkley Point C. (June). (https://nao.org.uk/wp-content/uploads/2017/06/Hinkley-Point-summary.pdf)

[17] Department for Business Energy & Industrial Strategy 2019 RAB model for nuclear: consultation on a RAB model for new (https://assets.publishing.service.gov.uk/government/uploads/system/uploads/attachment_data/file/825119/rab-model-for-nuclear-consultation.pdf)

[18] Helm D 2018 The Nuclear RAB Model (http://dhelm.co.uk/energy/energy/the-nuclear-rab-model/)

[19] Newbery D, Pollitt M, Reiner D and Taylor S 2019 Financing low-carbon generation in the UK: the hybrid RAB model. Vol. 1969. Cambridge Working Papers in Economics

[20] Puikkonen I 2010 Cooperative Mankala-companies—the acceptability of the company form in EC competition law *Helsinki Law Rev.* **4** 139–56

[21] Lindroos T J, Pursiheimo E, Sahlberg V and Tulkki V 2019 A techno-economic assessment of NuScale and DHR-400 reactors in a district heating and cooling grid *Energy Sources, Part B Econ Plan Policy* **14** 13–24 Available from: https://doi.org/10.1080/15567249.2019.1595223

[22] Helm D 2018 *Burn Out: The End Game for Fossil Fuels* Kindle edn (New Haven and London: Yale Univeristy Press) p 241

[23] UK Parliament 2019 Government gives details on setting a UK net zero emissions target News from Parliament UK Parliament (https://parliament.uk/business/news/2019/june/government-gives-details-on-setting-a-uk-net-zero-emissions-target/)

[24] Committtee on Climate Change 2019 Net zero technical report (www.theccc.org.uk/publications)

[25] UK Department for Business Energy and Industrial Strategy 2019 Digest of United Kingdom Energy Statistics (https://assets.publishing.service.gov.uk/government/uploads/system/uploads/attachment_data/file/840015/DUKES_2019_MASTER_COPY.pdf)

Section II

Progress in the private sector

IOP Publishing

Commercialising Fusion Energy
How small businesses are transforming big science
William J Nuttall, Satoshi Konishi, Shutaro Takeda and David Webbe-Wood

Chapter 5

Tokamak Energy

Melanie Windridge

Fusion has long held the promise of clean, safe, carbon-free, sustainable energy production. As described elsewhere in this book, over the past 10–15 years an increasing number of private ventures have entered the fusion space. As well as money, the private companies are bringing agility, innovation and a renewed drive to accelerate the development of commercial fusion power. Between them they are investigating a wide range of different approaches or new technologies.

As the private fusion industry gains maturity, fusion is becoming an increasingly compelling investment proposition. It has the societal angle—fusion could make a significant environmental impact while delivering flexible power on a global scale—but also a positive valuation trend looking to deliver very healthy returns to investors who can accept the long timescale in return for a high ratio of reward to risk. This is good news for fusion in general. More money will drive greater regulatory certainty, more scientific advances and a wider workforce, which will in turn drive more investment in a virtuous circle. However, to maintain and strengthen investor interest, it will be increasingly important that private ventures don't just tell a good story, they also deliver tangible, independently verifiable, scientific and technological milestones.

Tokamak Energy is one of the more established private fusion companies, founded in 2009 as a spin-out from Culham Centre for Fusion Energy, and having raised £117 m to date (July 2020). The company has established a team of over 110 employees and more than 60 contractors and consultants (July 2020), including numerous world-leading scientists and engineers. It continues to grow rapidly.

Tokamak Energy is building on mature physics with new technologies, effectively modernizing an established concept. The company strategy is to combine highly efficient 'spherical' tokamaks with high temperature superconducting (HTS) magnets. The team is confident that this combination will deliver fusion power *smaller, cheaper and faster*, and with a distinct competitive advantage over other approaches. Critically, Tokamak Energy is focused on the commercial goal. Fusion power needs

doi:10.1088/978-0-7503-2719-0ch5

to be commercially attractive in order to become rapidly deployable at scale. The claimed efficiency advantages of the spherical tokamak give Tokamak Energy a significant advantage in this respect.

Tokamak Energy has made impressive progress over the last ten years, operating a world-class tokamak device and breaking records in HTS magnet development. The company aims to demonstrate energy gain in their next device, a fusion power demonstrator named ST-F1 that is set to be built by 2025. The successor device, ST-E1, is targeted on demonstrating electricity to the grid by 2030.

This chapter will describe the work of Tokamak Energy towards delivering commercial fusion faster. We will consider the Tokamak Energy approach—including advantages of and evidence for—progress so far and plans for the future.

5.1 The case for fusion energy

Fusion has various attributes to commend itself to those looking to ensure reliable electricity generation to a population, see figure 5.1. Fusion produces energy from an abundant fuel supply with no carbon emissions arising from operations. Like fossil fuels, fusion is dispatchable, meaning that the power output can be adjusted to meet the demands of the population and is therefore able to provide reliable base-load power with the potential for load-following and/or co-generation of electricity and hydrogen. Fusion is safe—there is no risk of nuclear meltdown, no long-lived radioactive waste is produced and there is no fissile material.

		Weather independent	No CO_2	Abundant Supply / No Geographical Limitations	No Major Waste Concerns	Cost
Non-Fossil Fuel	Fusion Energy	✓	✓	✓	✓	✓
	Hydroelectric	✓	✓	✗	✓	✓
	Geothermal	✓	✓	✗	✓	✓
	Solar	✗	✓	✗	✓	✓
	Wind	✗	✓	✗	✓	Onshore ✓ Offshore ✗
	Nuclear Fission	✓	✓	✗	✗	✗
Fossil Fuel	Natural Gas	✓	✗	✗	✗	✓
	Coal	✓	✗	✗	✗	✗
	Oil	✓	✗	✗	✗	✗

Figure 5.1. Desirable attributes of power generation methods. Fusion is the only one that ticks all the boxes.

Private company approaches, such as Tokamak Energy's, involve fusion devices that are compact and efficient, and can be built at competitive cost. No other energy source ticks all these boxes.

5.2 The Tokamak Energy approach—spherical tokamaks and high temperature superconductors

Tokamak Energy offers a faster way to commercial fusion by overlaying new technologies on established physics. The company has pioneered the unique combination of spherical tokamaks (STs) and high temperature superconductors—a possibility that emerged only in the last decade since high temperature superconductor (HTS) materials became available as a reliable engineering material, fabricated into long tapes suitable for winding into magnets. This combination opens up the possibility of building smaller, more compact fusion devices than previously envisaged.

The Tokamak Energy plan is to develop smaller, lower-power modules producing ~150 MW_e, which may be grouped together to form a gigawatt-scale power plant [1] or operated on a distributed basis close to the places where the energy will be used. A modular design for future power plants would have the advantages of factory production (including cost reductions from economies of scale and economies of multiples) and more flexible plant operation.

Spherical tokamaks (STs) are low aspect ratio, meaning their shape is more squashed and compact than a conventional tokamak, rather like a cored apple rather than a ring doughnut. Two of the Tokamak Energy founders, Alan Sykes and Mikhail Gryaznevich, developed the concept in the UK in the 1990s at the Culham Centre for Fusion Energy. These experimental endeavours built upon theoretical work, such as that by Sykes [2] and Peng [3], which had indicated that the spherical tokamak (ST) shape could be more efficient.

Spherical tokamaks have some key advantages over conventional tokamaks:

- **Efficiency**—Their compact shape means that the plasma sits closer to the centre column where the magnetic field is generated. Therefore they make more efficient use of the magnetic field than conventionally-shaped tokamaks and can achieve higher β, the ratio of plasma pressure to magnetic pressure.
- **Improved confinement**—There are indications that spherical tokamaks exhibit better confinement than conventional tokamaks [4]. In other words, they can achieve a higher 'energy confinement time', or are better at keeping the plasma hotter for longer.
- **Stability**—Spherical tokamaks suffer less from performance-degrading instabilities. The compressed geometry means particles spend a greater portion of time on the inboard side of the plasma where the magnetic field is high and the field line curvature acts to stabilize instabilities driven by the pressure gradient. Additionally, ST plasmas are naturally more vertically stable at higher elongation [5]. Because of the magnetic field configuration and the proximity of the centre column, they naturally take the more efficient, elongated plasma shape without artificial stretching, which can make plasmas more susceptible to vertical instability.

- **Higher bootstrap current**—In spherical tokamaks, less energy is required for plasma stabilization via the plasma current due to the higher fraction of 'bootstrap current' that is achievable [6]. An electric current flowing through the plasma (i.e. the 'plasma current') is an important element of the tokamak, generating a poloidal magnetic field that provides plasma confinement along with the toroidal field. Plasma current is generally driven inductively by the central solenoid and for longer operation requires external methods such as neutral beams or radio frequency waves [7]. The bootstrap current is a toroidal current in the plasma generated by the interplay between trapped and passing particles in the presence of a density or temperature gradient. More information can be found in [6]. Wilson *et al* [8] derived an approximate scaling for bootstrap current fraction and showed that bootstrap fraction increases with normalized beta β_N and higher elongation, both of which are naturally higher in an ST. A higher bootstrap fraction means a lower requirement for an expensive and inefficient non-inductive current drive.
- **Less severe disruptions**—Evidence suggests that disruptions in spherical tokamaks produce less severe effects than those in conventional tokamaks, such as lower currents generated in the vessel walls—known as halo currents—and subsequently lower mechanical forces on the structure of the machine when these currents interact with local magnetic fields. Moreover, halo currents in spherical tokamaks are generally more uniform in distribution than those in conventional tokamaks, meaning that the tokamak will suffer less lateral loading on the vacuum vessel [9, 10].

High temperature superconductors (HTS) are a class of superconducting materials that become superconducting (i.e. lose all electrical resistance) at a higher temperature than the low temperature superconductors that have been widely used in applications like MRI and particle accelerators for many years. Conventional low temperature superconductors (LTS) operate around 4 K (−270 C). Second generation high temperature superconductors start to superconduct at around 90 K (−180 °C) [11], but cooling them further improves their performance. Operation at 20–40 K is optimal and still allows a huge saving in cryogenic cooling power compared to conventional 4 K operation. In principle it takes five times more energy to maintain a magnet at 4 K than at 20 K. In practice, in a real system, that ratio is even higher.

A key feature of high temperature superconductors is the ability to carry exceptionally high current density at high magnetic fields (well above 20 tesla), so they can generate a higher confinement field than conventional superconductors, and do so in a smaller device. Additionally, their construction means they take up less space, allowing designers to reduce centre column diameter and hence device size. Considerable engineering ingenuity is required to achieve the exceptionally high current densities in the centre column magnet, but this essential requirement drives innovation and the generation of intellectual property.

Tokamak Energy magnet designers are confident of achieving greater than 200 A mm^{-2} in the centre column of ST-F1 so that the millions of amps necessary to create the high fields can be carried in a centre column of around 200 mm radius. Validating this design is challenging, but Tokamak Energy has already built small test magnets that achieve over 1000 A mm^{-2} and magnetic fields on the conductor of over 24 tesla. The team is now building a much larger test magnet in a tokamak configuration with a centre column fully representative of the centre column required for ST-F1. This will have a centre column very close in size, current density, total current and compressive stress to ST-F1 and, importantly, it will validate the magnet quench protection system.

All superconducting magnets run some risk of 'quench' or sudden loss of superconductivity and must be designed to shut down safely in such an event. The risk of a quench can be minimized by design and it turns out that Tokamak Energy's designs produce HTS magnets that are difficult to quench. However, it is still necessary to prove that if a quench does happen the magnet will shut down safely, without being damaged.

More information on the advantages of spherical tokamaks and high temperature superconductors can be found in [12] along with some of the history of STs and tokamak development.

5.3 The combination of spherical tokamaks and high temperature superconductors

The combination of STs and HTS opens up the possibility of building smaller, cheaper, more efficient fusion devices, and of building them faster.

A hint of why this is possible is given in the expression

$$P \sim \beta^2 B^4 V$$

The fusion power produced increases as the square of β (the parameter in tokamaks that gives a measure of the efficiency of the machine), as the fourth power of B (the magnetic field at the centre of the plasma), and with V (the plasma volume, related to the machine size). Since β and B were thought to be the most difficult parameters to increase, historically fusion machines increased in size, notably from JET to ITER, which was a jump from 18 to 30 m high, and an increase in plasma volume from 80 to over 800 m^3.

Rather than increasing size, the Tokamak Energy plan is to increase the value of β by using a spherical tokamak design, while doubling the magnetic field strength, B, using high temperature superconductors. A conceptual design and solutions to various challenges are discussed in Sykes 2017 [13].

However, one should remember that the parameter that matters is not fusion power, P, itself, but fusion energy gain, Q_{fus}. The Tokamak Energy approach, supported by studies, is that the combination of the spherical tokamak geometry with the high magnetic field enables a design capable of achieving high fusion gain at modest size and modest fusion powers.

Costley explains this comprehensively in [14]. Q_{fus} is directly related to $nT\tau_E$, known as the 'fusion triple product' of density, temperature and energy confinement time. It is the figure of merit of a fusion plasma, against which progress towards fusion can be measured. $nT\tau_E$ must exceed 10^{21} m^{-3} keVs for net fusion power [6].

Costley states how it is possible to derive an expression for $nT\tau_E$ in terms of the main device parameters:

$$nT\tau_E \propto \frac{H^2}{q^3} R_0^2 B_T^3 \left(\frac{\kappa^{7/2}}{A^3} \right).$$

Here H is a numerical factor that allows for the fact that the actual confinement time found in experiments can deviate from scaling laws. It is typically in the range 0.8–2.0. The parameter q is the safety factor and is related to the pitch of the magnetic field in the plasma, which experiments show should be maintained at $\geqslant 2.5$. But what is important to note here is the dependence on size (plasma major radius R_0), field (B) and shape ($\kappa^{7/2}/A^3$, where κ is the plasma elongation and A is the aspect ratio calculated as major radius over minor radius, R_0/a).

The approach requires a magnetic field at the centre of the plasma of >3 T, which requires a greater field on the conductor itself due to the magnetic field dropping off as $1/R$, see figure 5.2. Tokamak Energy is considering a field of the order of 3–5 T for their fusion power demonstrator device ST-F1. While this is relatively low for a conventional tokamak (ITER will have magnetic field at the centre of the plasma of 5 T), it is significantly higher than that in other STs, something which Tokamak Energy is investigating in the ST40 tokamak, which is discussed in the next section. However, just because HTS *can* generate very high fields, and one naturally assumes the plasma will also be sitting in a higher field, that does not necessarily have to be the case. By adjusting the balance of size, field and shape parameters we find different design approaches to achieving the required fusion performance.

There are various design considerations related to the Costley expression above that give three different approaches for the same fusion gain:

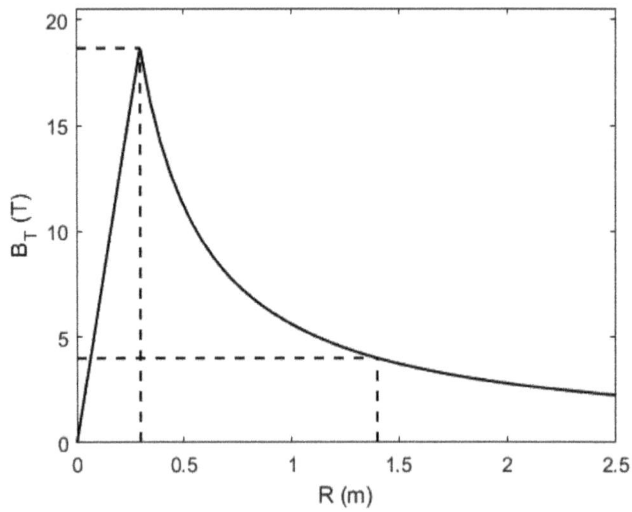

Figure 5.2. How the magnetic field changes with major radius, first increasing as R within the conductor and then dropping off as $1/R$ outside. This diagram illustrates the difference between the 'field on the conductor' which will need to be >20 T, and the field at the major radius of the plasma, which will need to be 3–5 T in ST-F1. Note that the conductor here would be only the current carrying part of the centre post, not including any shielding, which could shift it outwards.

- Conventional aspect ratio with LTS magnets (e.g. ITER): 'modest' field on coils, 'modest' field on plasma, large size for decent fusion gain.
- Conventional aspect ratio with HTS magnets (e.g. CFS approach): high field on coils, high field on plasma, allows smaller size for same fusion gain.
- Low aspect ratio ST with HTS magnets (e.g. TE approach): high field on coils, modest field on plasma, efficiency gains due to shape permit high fusion gain in small device.

It is not simply an improved efficiency that is an advantage of the spherical tokamak; it is the combination of factors discussed above, including the improved energy confinement and the increased bootstrap current fraction of the ST shape. These extra, ST-specific factors allow the possibility of reducing size without having to increase magnetic field excessively.

Finally, spherical tokamaks can make a substantial contribution to plant efficiency; they will be able to deliver net energy production through both a high fusion energy gain and low energy 'overheads'. The key overheads are current drive, required to sustain the plasma, and balance of plant. The spherical tokamak design can significantly reduce the external power required because the shape efficiency reduces plasma current by a third and, moreover, >96% of plasma current can be efficiently self-generated (the 'bootstrap current') [15] versus 60% in a conventional tokamak.

The spherical tokamak has long been seen as a candidate for an efficient fusion power plant if the engineering challenges of the more compact shape could be solved. The advent of high temperature superconductors is now making that possible.

5.4 Progress so far

Tokamak Energy has a good track record of achieving its milestones, and even exceeding them, particularly in the case of HTS magnets.

Tokamak Energy demonstrated initial capabilities with a small tokamak called ST25 ($R_0 = 0.25$ m). In 2015 the company showed the potential of high temperature superconducting materials as tokamak magnets in a similarly-sized machine with all magnets made of HTS. A plasma was held for 29 h in the ST25HTS and was an important early demonstration of the HTS magnet technology.

Following on from this milestone, Tokamak Energy built a larger machine, the world-class spherical tokamak ST40, whilst in parallel developing a 5T HTS demonstrator tokamak magnet.

> ST40 key parameters:
> $R_0 \sim 0.4$ m
> $R/a \sim 1.8$
> $I_p = 2$ MA (target)
> $B_{t0} = 3$ T (target)
> Copper magnets

ST40 was designed to meet the key milestone of achieving fusion temperatures (100 million degrees, or 10 keV) at densities ~1–5.10^{20} m^{-3}, but also to investigate how plasma behaves in a high-field, compact configuration and to research methods of solenoid-free start-up, particularly the technique of merging compression [16]. ST40 is the world's highest-field spherical tokamak, so is able to do important fundamental research in this unexplored parameter space. Tokamak Energy's interest is to test how plasma temperature, pressure and energy confinement time will vary as the toroidal field is increased. The results will influence the design of ST-F1 and subsequent fusion power modules. Some pictures of ST40 and the control room are shown in figure 5.3.

The importance of high magnetic field in a spherical tokamak has recently been emphasized by the US Department of Energy. They issued a call for proposals for US organizations wishing to receive DOE funding to work on Tokamak Energy's ST40. In the call for proposals DOE says:

> ST40 is a high field (3 T) spherical tokamak that has been built and is being operated by the private company Tokamak Energy in the UK. At 3 T, ST40 will operate at approximately three times the maximum operating toroidal field of NSTX-U. This world leading field strength provides a unique ST operating space, which can further broaden our knowledge of ST physics.

In May 2018, the team achieved temperatures of 15 million degrees in ST40, an important step on the way to 100 million degrees. After this successful campaign the company moved into a new, larger facility and ST40 was dismantled and re-built in its new home. ST40 has also undergone a series of upgrades since then, including adding further power supplies, installing neutral beam injectors (NBI) for plasma heating, adding a liquid nitrogen cooling system for the copper coils, and expanding the suite of diagnostics.

Once ST40 has achieved plasma temperatures of 100 million degrees (10 keV) there are plans to go on to demonstrate energy gain conditions.

Alongside the work on ST40, Tokamak Energy runs an HTS magnet development programme. The HTS team has worked with tapes from different suppliers to establish handling, construction and operational techniques on small magnet prototypes with a view to creating a full-scale tokamak magnet. The HTS programme is an integral part of the plan to deliver commercial fusion energy.

Very rapid developments were seen in 2018 to 2019 in magnet technology at Tokamak Energy, culminating in the team exceeding their 2020 target of 15 tesla with HTS magnets some 18 months early. Indeed, the magnet engineers succeeded in producing a magnetic field of 24.4 T in magnets made from tapes of the high temperature superconducting material REBCO (Rare earth–barium–copper oxide), see figure 5.4.

This conduction-cooled all-REBCO magnet achieved its peak field at 21 K in a cold bore of 50 mm, which is believed to be a record performance. In superconducting terms, 21 K is a relatively high temperature, so the cryogenic system is greatly simplified. Additionally, the magnet was shown to be extremely robust, reliable and simple to manufacture, with impressive defect tolerance and good response to a sudden loss of superconductivity, called a quench.

The progress of the HTS team has been faster than expected with milestones being hit well ahead of schedule (this milestone was scheduled for late 2020). Currently the team is working on scaling up these magnets into the configuration required for tokamaks.

Figure 5.3. Images of ST40 and the control room, 2019.

Figure 5.4. High temperature superconductor test coils developed as part of the HTS magnet development programme.

One advantage of private companies is their agility when faced with hurdles.

A good example of this was during the ST40 15 million degree campaign. The physics team decided that to optimize the shape and position of the plasma it would be better if they had a power supply that could deliver a different waveform, and an extra coil which could move the plasma slightly. On the face of it, these were both upgrades that would take many months, possibly years. Instead the team came up with some novel ideas for how to tackle the problem and went on to achieve their 15 million degree milestone. They proceeded to build, install and test solutions in a matter of weeks. This kind of entrepreneurial spirit and agility is a key attribute of private companies. Rather than accepting several months or years of delay due to hardware issues, the team was able to invent a way to do it quicker.

Videos of Tokamak Energy's plans and progress are available in a dedicated playlist on their YouTube channel.

5.5 Future plans and business overview

Tokamak Energy has formed a roadmap for the 'faster fusion' approach. The spherical tokamak and high temperature superconductor development tracks come together into ST-F1 and then ST-E1 (the 'F' and 'E' denoting 'fusion' and 'electricity' respectively).

ST-F1 will be a fusion power demonstrator—a spherical tokamak with all-HTS magnets capable of industrial scale heat production. Work is currently in progress on this device and it is planned for operation in 2025. ST-F1 will exceed energy breakeven conditions of $Q = 1$, demonstrating energy gain.

ST-E1 will be a commercial-scale power plant module capable of generating electricity to the grid. It will demonstrate the commercial viability of fusion energy ready for deployment during the 2030s.

The Tokamak Energy technology roadmap, figure 5.5, includes several key research areas on the path to commercialization. Some will be proprietary; most will be collaborative.

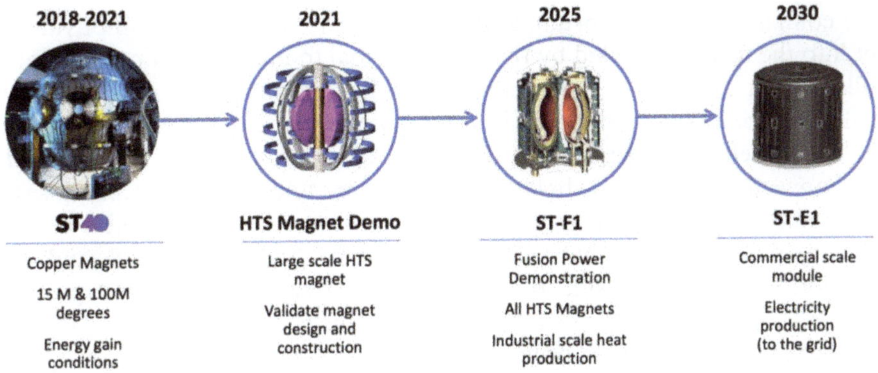

Figure 5.5. The Tokamak Energy roadmap where the development tracks of the ST40 tokamak and the HTS Magnet Demo come together into the tokamaks ST-F1 and ST-E1 to demonstrate first fusion power and then electricity to the grid.

Leading up to ST-F1 there will be research into auxiliary systems, pilot plant design, plasma control, heat and particle exhaust, radiation shielding and tritium fuel handling. Moving on to ST-E1 will require further work into tritium fuel breeding, energy harnessing technologies and advanced technology development as the company progresses from fusion power demonstration to full electricity generation.

There are too many challenges for Tokamak Energy to tackle them all in-house. Instead the strategy is to innovate through collaborations, often with universities and government laboratories. Tokamak Energy does develop specific technologies, such as HTS magnets, in-house, but only when a world-class team can be assembled and where there is an opportunity to generate valuable intellectual property (IP). In some cases in-house developments are necessary because there is insufficient external expertize.

As with any commercial enterprise, protecting IP is an important element of the business. Tokamak Energy has been innovating rapidly since its inception; IP is protected and continues to grow. However, the company understands that the range of challenges of fusion energy is too great for a single company to do everything alone. There is a balance to be struck between forming mutually-beneficial collaborations and generating IP. The company has collaborations with universities and organizations around the world, including Oxford, York and Imperial College in the UK, Princeton Plasma Physics Laboratory in the USA, the University of Tokyo in Japan, and CERN in Switzerland. Private companies can also benefit from cost-sharing programs such as the US Department of Energy's Innovation Network for Fusion Energy program, or INFUSE. These are designed to encourage private–public research partnerships to tackle challenges in fusion energy development by making accessible the huge wealth of technological expertize and long experience of US Department of Energy National Laboratories.

Tokamak Energy currently (July 2020) has over 50 separate patent families comprising 22 granted patents and over 80 pending applications. 35 of the patent families are around high temperature superconductors and cover applications beyond fusion.

The company has been fortunate in being able to attract highly-motivated key talent into the team, all united by the common goal to make fusion power available for all as soon as possible. Tokamak Energy has an impressive depth of technological expertize across its staff and its advisory boards: members of the engineering team have extensive experience of delivering major engineering projects and in-depth knowledge of state-of-the-art manufacturing and production processes; the HTS team is world-leading in the development, design and scale-up of superconducting magnets; others have expertize in control systems, materials and system integration. On the physics side, the company was founded by respected, pioneering physicists in the field of spherical tokamaks; eminent consultants have decades of experience from major fusion programs; and the younger members of the team are graduates from world-leading plasma and fusion science institutions.

With a solid plan for commercialization and a formidable team, Tokamak Energy is confident that it is in a good position to deliver on its aims.

5.6 Summary

The world has been waiting for fusion for a long time, but it could be argued that part of the reason was the external conditions and not simply the science itself. The science is, of course, extremely challenging, but now the coming together of mature fusion science, new enabling technologies, environmental considerations and private investment is altering the external conditions to a point where fusion has the opportunity to prove itself. Fusion is a disruptive game-changer coming sooner than many people might think.

Tokamak Energy believes it has a unique and winning plan in the combination of the spherical tokamak shape with high temperature superconducting magnets. The approach builds on established physics but brings in new technologies, private funding and private-sector agility.

The dedicated team is serious about delivery, as their track record attests.

The well-protected IP, including areas outside of fusion energy, drives valuation growth and gives investors confidence. And greater investment will drive progress and deliver fusion faster for all.

References

[1] Chuyanov V A and Gryaznevich M P 2017 Modular fusion power plant *Fusion Eng. Des.* **122** 238–52

[2] Sykes A, Turner M F, Fielding P J and Haas F A 1979 Plasma physics and controlled nuclear fusion research *Proc. 7th Int. Conf. 1978* Vol. 1 (Vienna: IAEA)

[3] Peng Y-K M and Strickler D J 1986 Features of spherical torus plasmas *Nucl. Fusion* **26** 769–77

[4] Ono M and Kaita R 2015 Recent progress on spherical torus research *Phys. Plasmas* **22** 040501

[5] Windridge M J 2009 Tokamak axisymmetric stability: vertical displacements and their consequences *PhD Thesis* chapter 5, section 5.4.3 pp 103–105

[6] Wesson J 2011 *Tokamaks* 4th edn (Oxford: Oxford University Press)

[7] Westerhof E 2012 Non-inductive current drive *Fusion Sci. Technol.* **61** 312–9

[8] Wilson H R *et al* 2003 FT/1-5 the spherical tokamak fusion power plant *19th IAEA Fusion Energy Conf. (Lyon, France)*
[9] Counsell G F, Martin R, Pinfold T, Taylor D and the MAST team 2007 On the magnitude and distribution of halo currents during disruptions on MAST *Plasma Phys. Control Fusion* **49** 435–46
[10] Eidietis N W, Gerhardt S P, Granetz R S, Kawano Y, Lehnen M, Lister J B, Pautasso G, Riccardo V, Tanna R L and Thornton A J 2015 and The ITPA disruption database participants. The ITPA disruption database *Nucl. Fusion* **55** 063030
[11] Cardwell D A and Ginley D S 2003 *Handbook of Superconducting Materials* Vol 1 (Bristol: Institute of Physics Publishing)
[12] Windridge M 2019 Smaller and quicker with spherical tokamaks and high-temperature superconductors *Phil. Trans. R. Soc.* A **377** 20170438
[13] Sykes A *et al* 2017 Compact fusion energy based on the spherical tokamak *Nucl. Fusion* **58** 016039
[14] Costley A E 2019 Towards a compact spherical tokamak fusion pilot plant *Phil. Trans. R. Soc.* A **377** 20170439
[15] Andrade M C and Ludwig G O 2008 Scaling of bootstrap current on equilibrium and plasma profile parameters in tokamak plasmas *Plasma Phys. Control Fusion* **50** 065001
[16] Gryaznevich M P and Sykes A 2017 Merging-compression formation of high temperature tokamak plasma *Nucl. Fusion* **57** 072003

IOP Publishing

Commercialising Fusion Energy
How small businesses are transforming big science
William J Nuttall, Satoshi Konishi, Shutaro Takeda and David Webbe-Wood

Chapter 6

Laser fusion CANDY GPI/Hamamatsu

Y Mori

6.1 Introduction

Research and development of fusion power has been on-going since the 1950s to realize it as a safe, unlimited source of energy that does not emit any carbon dioxide. Fusion power, which includes conversion of mass into energy through fusion of nuclei, is the principle behind power generation in the Sun. All energy resources on our world have been energized by the Sun since the birth of life billions of years ago. The Sun powers the recirculation of the oceans, the atmosphere, hydropower, wind-power, solar power, and fossil fuels which originated from photosynthesis of ancient plants. Especially, fossil fuels energized by the Sun support modern society as the primary source of energy. With the catch phrase, ‘Sun on the Earth’, the efforts to realize controlled fusion power as a future energy resource started in the 1950s. There are three main highlights of fusion power: it is safe, it is an inexhaustible resource, and it does not emit carbon dioxide.

Research and development efforts have focused on generating fusion power that exceeds input power. Fusion power requires external input energy to fuse the nucleus of the fuels through burning at very high temperature. The achievement of controlled fusion power close to the burning condition was demonstrated in the 1990s by confining fusion fuels into gaseous plasma in the magnetic field basket [1]. This success led to the international project ITER [2] aimed at burning fusion fuel in order to generate fusion power exceeding ten times the input energy. It is currently under construction and is expected to be functioning by 2035.

The burning of fusion fuel by confining solid fuels in a high-density (~1000 times the solid density) state with laser-driven instantaneous pressure is also being conducted at the National Ignition Facility (NIF) in the US [3].

Both ITER and NIF are large-scale projects designed primarily to realize generation of fusion power that exceeds input power. The efforts to generate fusion power and convert it into electricity are considered as the next step after the achievement of fusion burning.

doi:10.1088/978-0-7503-2719-0ch6

We proposed a laser fusion mini-reactor, CANDY [4], that demonstrates fusion energy extraction at no-gain condition: fusion power is much lower than the input power. The concept of CANDY is to include a major component towards future power plant for the elimination of technical hurdles. Especially, recent progress in solid state semiconductor technology has enabled repetitive operation of a high-power laser system scalable to future laser fusion power plants. This chapter describes the progress of CANDY. Following this introduction, laser fusion activities around the world are described in section 6.2. Section 6.3 describes laser fusion activities in Hamamatsu by featuring the background to our activities and a strategy leading to CANDY. Section 6.4 describes the progress of CANDY by focusing on three key issues: (i) laser driver, (ii) plasma fuel physics, and (iii) target injection and engagement. Section 6.5 summarizes the chapter.

6.2 Laser fusion activities in the world

6.2.1 How the laser fusion reactor generates power

Burning of fusion fuel by laser pulse requires a high-density condition of 600 billion atmospheres, more than twice the pressure condition at the centre of the Sun, within a confinement time of 100 ps. The criteria of power production at fusion power plants were first derived by J D Lawson [5]. Using the balance of power loss and fusion power, he derived the so-called Lawson criteria, which states that the product of fuel density with fuel confinement time should be greater than 10^{14} cm^{-3} s with fuel temperature beyond 5 keV for hydrogen isotopes of deuterium and tritium (DT), which react most easily on the Earth. This condition can be revised into the product of pressure and the confinement time; 10 atm s, by considering fuel temperature. There are two approaches or confinement schemes to satisfy this condition, as shown in figure 6.1(a). One is magnetic confinements where gaseous plasma is confined for a long time for steady state, and the other is inertial confinement where dense plasma is confined with instantaneous time. Laser fusion is an inertial confinement scheme that requires confining the fuel to 1000 times of its solid density with confinement time of 100 ps. The instantaneous fuel pressure then reaches 600 billion atm that is more than two times that at the centre of the Sun with 240 billion atm. To induce such a highly dense and high-temperature condition, a fuel capsule with an inner layer of DT ice is pushed using pressure created by illumination of laser or laser induced x-ray; this process is called implosion.

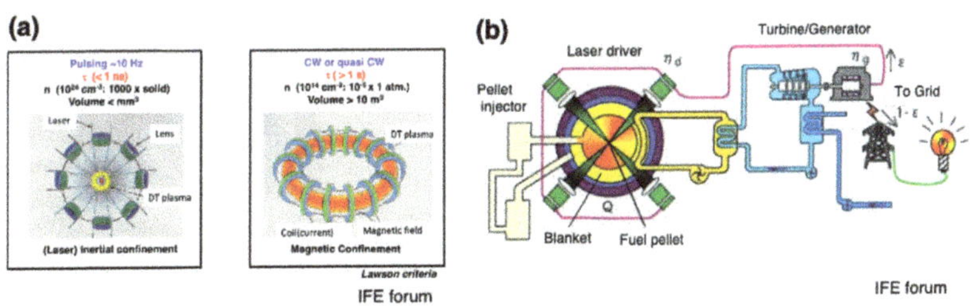

Figure 6.1. (a) Two approaches for fuel confinement. (b) Schematics of laser fusion power plant.

Laser fusion reactors require a five cycle process: capsule injection, laser engagement, fusing the fuel, capturing energy/conversion, and reactor evacuation. Figure 6.1(b) shows schematics of a laser fusion power plant. First, the fuel capsule is injected into an evacuated reactor. When the capsule reaches the centre of the reactor, lasers are engaged/illuminated on the surface of the capsule and deposit their energy on the capsule surface. The absorbed laser energy is converted to a high pressure to implode the fuel into a highly dense and high-temperature state, resulting in the ignition of compressed fuel. Fusion reaction power is absorbed at the chamber wall where liquid lithium lead (Li–Pb) is circulated. This liquid lithium lead flow captures x-rays/gamma-rays, charged particles, debris and fusion induced neutron energy, which is then converted into heat energy followed by conversion to socially useful energy forms such as electricity. The reactor requires a recovery time to evacuate the chamber ready for the next shot. For a steady state power plant, the above cycle needs to be repeated.

A commercial laser fusion reactor requires energy gains beyond 100 at an operation cycle of 10–30 Hz. In general terms, the fusion power plant is an electricity amplification machine, achieved by consuming electricity. For a laser fusion power plant, we should consider two criteria: energy gain (output fusion energy divided by the input laser energy) and economic efficiency (which depends on operation cycle frequency). For the energy gain calculated following Lawson's approach, if we consider the laser driver electricity efficiency of 10%, heat-electricity conversion efficiency of 33%, and recirculation ratio of electricity into the laser driver to be 25%, then the required energy gain is evaluated to be 120. For economic efficiency, the electricity for the grid is proportional to the product of output fusion energy with the number of operation cycle. Considering the reactor capital cost, a utilization efficiency of the reactor should be high. Several laser fusion reactor designs set the operation cycle at 10–30 Hz.

6.2.2 Laser fusion activities and trends

Laser fusion activities around the world are focused on break-even, where fusion energy exceeds the input laser energy. They utilize the so-called single-shot laser driver that provides more than 10 kJ laser energy pulse per shot with conventional laser technology and glass laser medium with flush lamp pumping. NIF in the USA is the leader in the single-shot laser driver technique. NIF injected a 1.8 MJ pulse into the fusion capsule and demonstrated fusion energy beyond 50 kJ. The repetition of such giant laser shots is limited by the cooling of the laser medium, which requires at least 3 h per shot. Figure 6.2(a) shows laser driver energy corresponding to the repetition. Single-shot laser drivers are installed in Japan, the UK, the US, France, and China. These facilities are working on understanding the plasma fuel physics needed for increasing the energy gain beyond unity.

Hamamatsu are working on a cycle operation of laser fusion experiments to pave the way for future power plants. For a laser fusion power plant, we need to modify the laser driver for repeated operation from a single shot. Figure 6.2(b) shows the strategy of the Hamamatsu laser fusion programme with respect to the world trend.

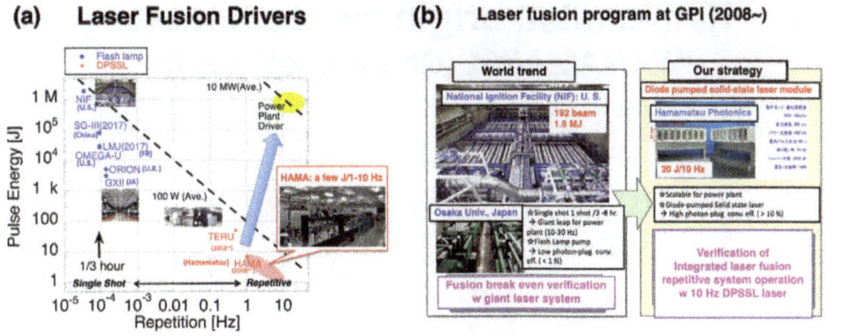

Figure 6.2. (a) Laser driver energy with respect to repetition. (b) Strategy of Hamamatsu laser fusion programme.

World trend dictates break-even of fusion plant utilizing a giant laser system. Our strategy is to utilize a high-cycle laser driver for verification of an integrated laser fusion high-cycle operation system that paves the way for future power plants. The laser driver for future power plants requires photon-wall-plug conversion efficiency beyond 10% and operation cycle beyond 10 Hz. A semiconductor diode-pumped solid state laser is one such candidate for the laser fusion driver. We developed the high-cycle laser fusion driver HAMA (Joule-class, 1–10 Hz) [6] as shown in figure 6.2(a) and demonstrated laser induced fusion reaction [7].

6.3 Laser fusion activities in Hamamatsu

6.3.1 Why laser fusion in Hamamatsu?

The laser fusion programme in Hamamatsu was initiated by Teruo Hiruma, honorary chairman of Hamamatsu Photonics, founder of the Graduate School for the Creation of New Photonics Industries (GPI). He envisioned producing electricity at a cost of 3.5 yen per kWh by laser fusion [8]. This electricity cost is comparable to that in Canada. He was passionate about maintaining industrial competency in Japan, where the cost of electricity is as high as 14 yen per kWh. Since the 1990s, he has encouraged colleagues at Hamamatsu Photonics to develop a semiconductor diode-pumped solid state laser that is one of the key technologies for future laser fusion power plants.

Hamamatsu, where Teruo Hiruma grew up, is full of challenge spirits. Hamamatsu is located at roughly the same distance from both Tokyo and Osaka along Tokai-do, one of the main roads in Japan. The Hamamatsu area is the headquarter of entrepreneurs who founded famous enterprises in Japan. Figure 6.3 shows the birth-places of the entrepreneurs hailing from Hamamatsu and GPI. Many entrepreneurs who represented Japanese companies grew up around the Hamamatsu/Hamana-lake area: Sakichi Toyoda (1867–1930) [9], the founder of Toyota, Torakusu Yamaha (1851–1916) [10], the founder of Yamaha, Soichiro Honda (1906–91) [11], the founder of Honda, Koichi Kawai (1886–1955) [12], the founder of Kawai, Michio Suzuki (1887–1982) [13], the founder of Suzuki, and Kenjiro Takayanagi (1899–1990) [14], a pioneer in television technology. There is a

Figure 6.3. The original Hamamatsu/Hamana-lake area entrepreneurs.

reason why many entrepreneurs hail from Hamamatsu. In Hamamatsu, there is a culture called 'YARAMAIKA' that in the dialect of this area, means 'challenge'.

The city of Hamamatsu features photonics to promote both photonics science and photonics industries collaborating with both academia and industry. Hamamatsu Photonics [15] is one of the key organizations. With 'YARAMAIKA' spirit, Heihachiro Horiuchi, a former student of Kenjiro Takayanagi, founded Hamamatsu Photonics in 1948. Teruo Hiruma, one of the board members of this company since the start-up phase, succeeded as the second president in 1978, and developed the photonics industries in the Hamamatsu area. As the third president since 2009, Akira Hiruma has been promoting the photonics industry deep in the Hamamatsu area. Following these activities for establishing Hamamatsu as a pre-eminent photonics city, the Photonics Declaration 2013 was signed [16].

Teruo Hiruma founded GPI in Hamamatsu with the purpose of education and promotion of new photonics industries in 2005. The vision of GPI constitutes a contribution to the healthy development of civilization and humanity on Earth, the dignity of our nation, happiness of the citizens, and activating the nation with Photonics [17]. To realize this vision, the GPI now covers six fields; Photonics Industry Management, Photonics for Material Processing, Optical Information and Systems, Photonics for Energy, Biophotonics, and Medical Photonics. The GPI aims to train PhD students with an 'entrepreneur' mentality through practice.

Teruo Hiruma started promoting the laser fusion programme in GPI in 2005 to realize his dream of generating electricity by laser fusion. Professor Yoneyoshi Kitagawa has conducted a laser fusion programme in GPI in collaboration with Hamamatsu Photonics, and the Toyota motor Corporations since 2008.

6.3.2 Strategy of GPI

The strategy of GPI is based on a repetitive laser system scalable to future laser fusion energy. The key technology is the semiconductor diode-pumped solid state laser system. There are three phases toward the adoption of laser fusion reactors.

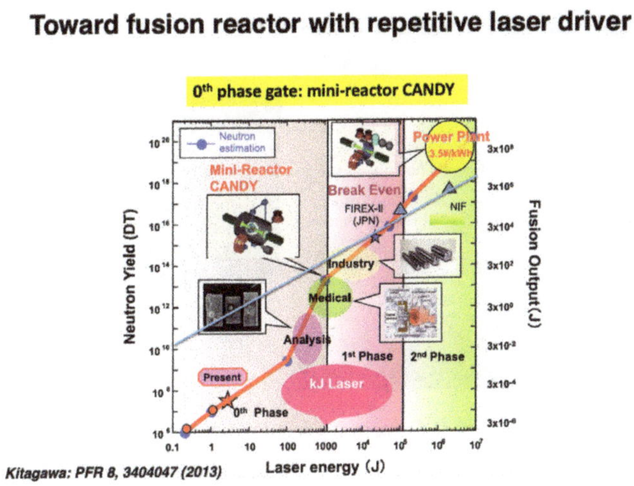

Figure 6.4. Neutron yield versus input laser energy: phase zero (0.1–1 kJ laser, 1 × compression), phase one (1~100 kJ, 100 × compression) and phase two (100 kJ–10 MJ laser, 100–500 × compression), Kitagawa PFR 2013. Reprinted from [18] with permission of JSPF.

Figure 6.4 shows neutron yield or fusion output energy, with respect to the laser energy [18]. On the roadmap, there are three phases: zero, one and two. Phase zero is the integrated mini-reactor, phase one is break-even, and phase two is developing a power plant. The phase gate of phase zero is construction of CANDY that utilizes the kJ-class laser system to generate 10 J-class fusion energy.

6.4 Laser fusion mini-reactor CANDY

6.4.1 Concept of CANDY

CANDY is proposed to demonstrate energy extraction from fusion reactions. Figure 6.5 shows an illustration of CANDY, which is composed of two parts: production of fusion reaction and collection of fusion energy. Fusion reactions are induced in the chamber centre. Fusion energies are collected through liquid metal flow that covers the inner wall of the reactor.

CANDY includes major compoenents for future power plants to eliminate the technical hurdles of the reactor. Cryo-tritium–deuterium-pellets are injected into the chamber centre at 10 Hz. Laser pulses are illuminated into each pellet from two sides to implode and ignite the fuel at 10 Hz. The inner wall of the reactor is covered by liquid lithium lead metal, the so-called blanket. This liquid metal blanket captures fusion energies during the reaction. The liquid metal circulates and carries the heat outside the reactor to a heat exchanger, which is connected to an electricity generator. The liquid metal also produces tritium; one of the elements of the fusion fuel. CANDY is designed with 4 kJ/10 Hz (40 kW) laser system and expected to deliver 200 W fusion power [19]. In CANDY, we focus on three key issues: (i) laser technology, (ii) plasma fuel physics, and (iii) target injection.

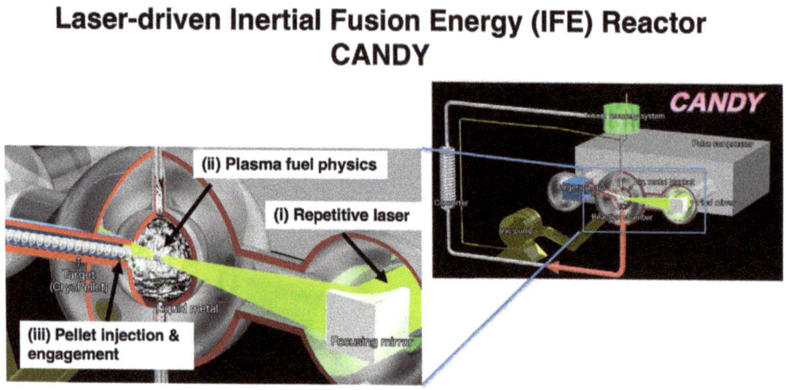

Figure 6.5. Laser fusion mini-reactor CANDY.

6.4.2 Laser technology

The semiconductor diode-pumped solid state laser (DPSSL) is one of the candidates for driving future laser fusion reactors. Research and development of high-power diode-pumped solid state laser has been conducted in Japan (HALNA laser: 20 J/10 Hz) and the U.S. (MERCURY: 50 J/10 Hz) since the late 1990s aiming to increase the photon-to-wall-plug efficiency beyond 10%. These laser systems are developed as one of the key equipments of laser fusion system, in separation from integrated laser fusion experiments.

The HAMA laser is the first repetitive Inertial Confinement Fusion (ICF) laser system utilizing DPSSL to demonstrate fusion reaction. Following the strategy of the Hamamatsu laser fusion project based on the repetitive laser system scalable to future laser fusion energy, we applied semiconductor DPPSL laser for laser fusion experiments [7]. Figure 6.6 shows the HAMA laser system. The HAMA laser applied DPPSL as the pumping source for an ultra-intense laser system [6] utilizing the so-called chirped amplification (CPA) system that enables laser fusion reaction with the J-class, table-top laser system. This CPA technology is one of the innovative technologies in laser physics, for which the Nobel prize in physics was awarded in 2018 [20].

Currently, development of DPPSL is on-going with respect to building beam lines in Hamamatsu. Hamamatsu Photonics has completed a new beam line for laser fusion experiments that allows measurement of laser induced pressures up to one million atmospheres [21]. In the laser irradiation building constructed in 2014, researchers have been conducting basic research into laser fusion plasma using a beam line consisting of a 100 J-class laser, laser beam transmission pipes, and a large spherical chamber. This new beam line will accelerate research on high-repetition rate laser fusion.

6.4.3 Plasma fuel physics

We applied the fast ignition (FI) scheme, which is an advanced ignition design. Figure 6.7(a) shows the ignition scheme of laser fusion. In the conventional ignition

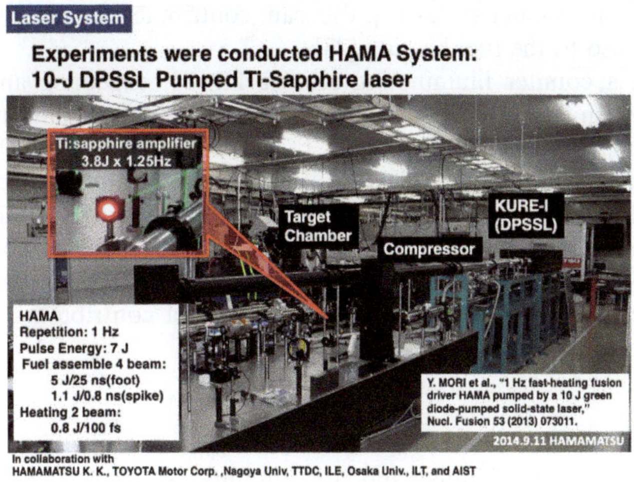

Figure 6.6. The HAMA laser system.

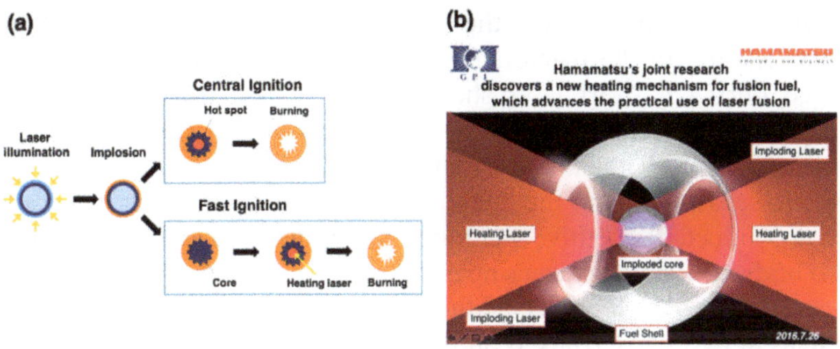

Figure 6.7. (a) Laser fusion ignition scheme, (1) central ignition and (2) fast ignition. (b) Illustration of counter illuminating fast ignition scheme.

scheme, the so-called 'central ignition', fusion fuel is imploded and ignited like in a diesel engine. In order to ignite the fuel, a hot spot must be generated at the centre of the imploded fuel. To produce a hot spot, compression symmetry of less than 5% is required. This is a challenging issue, in which the NIF is now involved. In the FI scheme, the processes of implosion and ignition are separated like in the gasoline engine. The hot spot is generated by illuminating an ultra-intense laser pulse into the imploded fuel. This separation relaxes compression symmetry for ignition. The original concept of FI was proposed in 1983. Invention of laser technology, CPA, enabled this advanced concept.

FI is a promising scheme to heat the imploded fuel and to enhance fusion reaction. We have demonstrated an ion temperature of 20 million degrees, which is hotter than the core of the Sun. This was achieved using the high-power laser system GXII/LFEX at Osaka University, in collaboration with the Hamamatsu team and the Osaka team [22]. The enhancement of fusion reactions reveals that fast ions produced from ultra-intense laser pulse can contribute not only to the fusion reactions, but also to the fuel heating [23].

We applied a counter illuminating scheme for FI, expecting an efficient fuel heating mechanism. Figure 6.7(b) illustrates the heating mechanism for counter illuminating the FI scheme applied by us. In the design of the CANDY reactor, we applied a counter illuminating scheme as start-up from minimum laser illuminating configuration to convert momentum energy into internal energy. On the laser fusion experiments using HAMA, we found a new heating mechanism: counter electron flows driven by a laser can drive the strong magnetic field inside of the core [24, 25]. This magnetic field can trap electrons inside the fuel contributing to the energy deposition into the fuel.

6.4.4 Target injection

Target deliveries into the chamber and laser illumination are the key issues to realize a steady state laser fusion reactor. In the laser fusion experiments so far, the shot rate of laser driver is limited for a few hours per shot. Therefore, there is enough time for target setup and alignment before the shot. In contrast, for high-cycle laser fusion experiments leading to the reactor, the target should be manufactured, delivered, and aligned during a short laser shot interval of less than 1 s. Research and development of the target system for future laser fusion power plant are being conducted in the US, Russia, and Japan. These activities are focusing on element development, separated from the integration fusion experiments.

We have conducted repetitive integrated target delivery and illumination experiments. To bridge the gap between single shot and repetitive laser fusion experiments, we have conducted development of the target supply systems with respect to the pellet designs in two steps: (i) fixed on the holder and (ii) injector without supporter [26]. The first target design is a spherical bead fabricated from deuterated polystyrene with a diameter of 1 mm applied for neutron generation. The first feeding system was a circular disk that supported beads. To achieve the configuration of an IFE reactor, the feeding system was switched to an injection system. Using a free-fall injection system, the engagement of 1 Hz-injected flying pellets involving fusion neutron reaction has been demonstrated as shown in figure 6.8(a) [27, 28].

This free-fall injection system is upgraded to 10 Hz, the same frequency as the laser operation. Figure 6.8(b) shows photograph of the 10 Hz injector installed in the illumination chamber. After five years of operation of the 1 Hz beads injection system [29], the system was upgraded to 10 Hz, the same frequency as the laser operation. Fuel pellets stored in the 'pellet loader' are distributed into 200 holes on the periphery of the 'rotating disk', where fuel pellets are injected at 10 Hz to the 'laser beam' engagement position. Then the two counter 'laser beams' hit the pellet at 10 Hz. This is the key technology for the laser fusion power plant. This is the first experimental demonstration of 10 Hz repetition injection of fuel pellets and induces fusion reaction advancing one step to realize a small-scale fusion reactor [30].

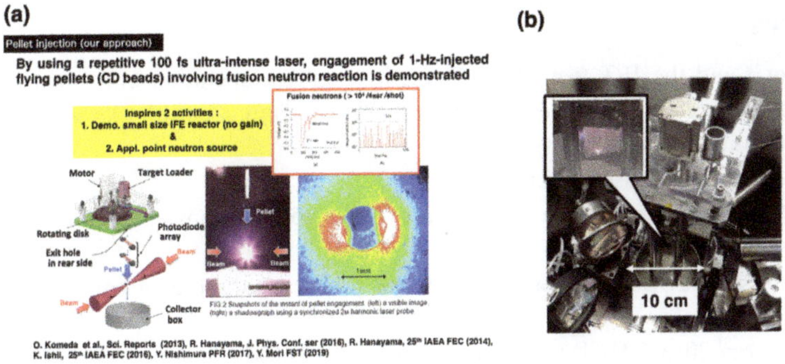

Figure 6.8. (a) Engagement of 1 Hz-injected flying pellets involving fusion neutron reaction, and (b) photograph of the 10 Hz injector installed in the illumination chamber.

6.5 Summary

We summarized the laser fusion activities in GPI, Hamamatsu including the background of the project. In this project, with collaborators mainly from the private sector, we proposed the laser fusion mini-reactor CANDY, which demonstrates fusion energy extraction in the condition of fusion power less than input power. CANDY includes major components for future power plants to clarify technical hurdles of the reactor. In CANDY, we focused on three key issues: (i) laser technology, (ii) plasma fuel physics, and (iii) target injection. The concepts of CANDY can contribute to the social acceptance of fusion power and accelerate the fusion programme.

Acknowledgments

This project was conducted with collaboration between the Graduate School for the Creation of New Photonics Industries, Hamamatsu Photonics K K, the Advanced Material Engineering Division of the Toyota Motor Corporation, TOYOTA Central Research and Development Laboratories, Green Mobility Research Institute, Nagoya University, Toyota Technical Development Corp, Purdue University, Centre for Materials Under Extreme Environment (CMUXE), Institute for Laser Technology, Institute for Laser Engineering, Osaka university, National Institute of Advanced Industrial Science and Technology, National Institute for Fusion Science, and Hiroshima university. The author acknowledges for project members and supporters; Y Kitagawa, K Ishii, R Hanayama, S Okihara, K Fujita, Y Nishimura, O Komeda, S Nakayama, S Suita, Y Takiguchi, Y Kato, S Nakai, T Sekine, R Yasuhara, T Katai, Y Takeuchi, T Kurita, T Watari, N Satoh, Y Hatano, R Yoshimura, M Takagi, Y Kabeya, H Kimura, Y Mizuta, Y Katoh, Y Tamaoki, N Kurita, H Kan, T Kawashima, O Komeda, N Nakamura, Y Miyamoto, T Kondo, M Fujine, S Tsuji, J Okishima, H Suto, Y Umetani, H Hirata, T Nishi, M Kakeno, H Azuma, T Hioki, T Motohiro, A Sunahara, Y Sentoku, T Nagai, Y Abe, Y Arikawa, S Fujioka, E Miura, A Iwamoto, T Ozaki, H Sakagami, and T Johzaki. We would like to thank Editage (http://www.editage.com) for English language editing.

References

[1] Jaquinot J and the JET team 1999 Deuterium–tritium operation in magnetic confinement experiments: results and underlying physics *Plasma Phys. Control. Fusion* **41** A13

[2] ITER website: https://www.iter.org

[3] NIF website https://lasers.llnl.gov/

[4] CANDY Introduction Movie (5 min.) http://www.youtube.com/watch?v=lTGgY8ZFR6Y&feature=youtu.be

[5] Lawson J D 1957 Some criteria for a power producing thermonuclear reactor *Proc. Phys. Soc.* B **70** 6–10

[6] Mori Y *et al* 2013 1 Hz fast-heating fusion driver HAMA pumped by a 10 J green diode-pumped solid-state laser *Nucl. Fusion* **53** 073011

[7] Kitagawa Y *et al* 2011 Efficient generation of DD fusion by all diode-pumped solid-state laser *Plasma Fusion Res.* **6** 1306006

[8] Hiruma T 2003 *Dekinai to Iwazuni Yattemiro!* (Japan: East Press) in Japanese

[9] Sakichi Toyoda, Wikipedia https://en.wikipedia.org/wiki/Sakichi_Toyoda

[10] Torakusu Yamaha, Wikipedia https://en.wikipedia.org/wiki/Torakusu_Yamaha

[11] Soichiro Honda, Wikipedia https://en.wikipedia.org/wiki/Soichiro_Honda

[12] Koichi Kawai, KAWAI web page: Founder/Inventive Genius https://kawaius.com/company/koichi-kawai/

[13] Michio Suzuki, Wikipedia https://en.wikipedia.org/wiki/Michio_Suzuki_(inventor)

[14] Kenjiro Takayanagi, Wikipedia https://en.wikipedia.org/wiki/Kenjiro_Takayanagi

[15] *Hamamatsu Photonics: Our Philosophy* https://www.hamamatsu.com/jp/en/our-company/our-philosophy/index.html

[16] Shizuoka University Press Release: *Photonics Declaration 2013 Signed in Hamamatsu, Japan* https://www.shizuoka.ac.jp/news/2013/20130611_eng.pdf

[17] GPI Web Page: *Founding Spirits* https://www.gpi.ac.jp/english/spirit/

[18] Kitagawa Y *et al* 2013 Hi-rep. counter-illumination fast ignition scheme fusion *Plasma Fusion Res.* **8** 3404047

[19] Kitagawa Y *et al* 2016 Progress toward a unified kJ-machine CANDY *J. Phys. Conf. Ser.* **688** 012049

[20] *Scientific Background on the Nobel Prize in Physics 2018* https://www.nobelprize.org/uploads/2018/10/advanced-physicsprize2018.pdf

[21] Hamamatsu Photonics Press Release: *Hamamatsu Photonics Has Completed a New Beam Line for Laser Fusion Experiments That Allows Measurement of Laser-induced Pressures up to One Million Atmospheres* https://www.hamamatsu.com/resources/pdf/news/2018_10_18_en.pdf

[22] Osaka University Press Release: *Success in Heating Fusion Fuel to 20 Million Degrees, Hotter Than the Core of the Sun, Using LFEX, World's Largest Laser* https://resou.osaka-u.ac.jp/en/research/2015/20150506_1

[23] Kitagawa Y *et al* 2015 Direct heating of a laser-imploded core by ultraintense laser-driven ions *Phys. Rev. Lett.* **114** 195002

[24] Mori Y *et al* 2016 Fast heating of imploded core with counterbeam configuration *Phys. Rev. Lett.* **117** 055001

[25] Mori Y *et al* 2017 Fast heating of fuel assembled in a spherical deuterated polystyrene shell target by counter-irradiating tailored laser pulses delivered by a HAMA 1 Hz ICF driver *Nucl. Fusion* **57** 116031

[26] Mori Y *et al* 2019 A spherical shell pellet injection system for repetitive laser engagement *Nucl. Fusion* **59** 096022

[27] Komeda O *et al* 2013 First demonstration of laser engagement of 1-Hz-injected flying pellets and neutron generation *Sci. Rep.* **3** 2561

[28] Nishimura Y *et al* 2017 Repetitive laser fusion experiment and operation using a target injection system *Plasma Fusion Res.* **93** 421–7 http://www.jspf.or.jp/Journal/PDF_JSPF/jspf2017_09/jspf2017_09-421.pdf

[29] Mori Y *et al* 2019 1-Hz bead-pellet injection system for fusion reaction engaged by a laser HAMA using ultra-intense counter beams *Fusion Sci. Technol.* **75** 36–48

[30] GPI Press Release: *Succeeded in 10 Hz Repetition Injection of Fuel Pellets* https://www.gpi.ac.jp/media/presentation.pdf

IOP Publishing

Commercialising Fusion Energy
How small businesses are transforming big science
William J Nuttall, Satoshi Konishi, Shutaro Takeda and David Webbe-Wood

Chapter 7

Pioneers of commercial fusion

Richard J Pearson and William J Nuttall

7.1 Introduction

Early scientific successes with the tokamak-type fusion reactor resulted in its eventual establishment at the centre of civil fusion R&D focus. Over time, support for the tokamak from governments around the world resulted in emphasis on the ITER project. However, it was not just tokamaks but other magnetic confinement fusion (MCF) approaches, such as the stellarator, that were also in focus in government-funded experimental programmes. Such approaches were seen as close competitors or as pathways that could provide technology development support for tokamaks. Inertial Confinement Fusion (ICF) has also been well supported in some countries, in part because of its links to research into nuclear weapons. ICF activities, in particular in the US (NIF at LLNL), in the UK (the Orion laser at the Atomic Weapons Establishment), and in France (the Laser Mégajoule at the Commissariat à l'Énergie Atomique), have seen substantial government support. A portion of the activities in such programmes is being geared towards supporting civilian science and engineering alongside the core mission of defence.

Since the turn of the millennium, there has been increasing private investment into fusion. Perhaps naturally, the most well-funded private sector companies are also focused on MCF reactor concepts, and mostly these have spun out of successful government-funded programmes. Notable examples are Commonwealth Fusion Systems and Tokamak Energy, both of which are pursuing routes to commercial fusion based on the tokamak. Comparably few companies are focused on ICF using lasers, the typical configuration for an ICF reactor (see chapter 2). Others are exploiting less well-known and more exotic MCF and ICF concepts, including in—arguably the oldest fusion concept of all—inertial electrostatic confinement (discussed in chapter 11). By way of example, we note the efforts of First Light Fusion and Lockheed Martin's Skunk Works division as private entities in pursuit of alternative ICF and MCF fusion approaches, respectively. Surprisingly, however, the majority of companies are not pursuing concepts that sit in either the MCF or ICF spaces at all, but rather are somewhere in the space in between.

As detailed more fully in chapter 2, MCF and ICF approaches can broadly be distinguished as having low-density plasmas with long confinement times and high-density plasmas with short confinement times, respectively. The space between these two approaches, which combines some aspects of both, is a conceptual space known as magneto-inertial fusion (MIF). Concepts in the MIF space have not been given the same attention as MCF, as they did not demonstrate the same early progress as the tokamak, and nor did they have any relevance to weapons technology that underpinned much of the drive behind ICF research. Consequently, and by comparison, MIF conceptual approaches received little support, including by way of public funding. However, MIF as well as other more exotic concepts have not arisen from nothing. Most MIF concepts that are now under development funded by private money also have their roots in fusion research programmes. However, such programmes were only ever in the background at laboratories and universities, operating on relatively small budgets and with small teams of scientists. The age of private investment into fusion is upending this.

So, why an ostensibly sudden interest in MIF? As mentioned, the MIF space is littered with many previously side-lined approaches that were developed in previous decades as part of early public fusion research programmes. The interest, per se, in MIF is thus not new, but rather 'renewed'. Much of the rationale for supporting MIF approaches now is that many of the reactor designs under development show the potential for low cost and simplicity—key criteria for a commercially viable fusion reactor. This is especially true when compared with the incumbent option under development in the public fusion programmes—especially large tokamaks. As such, MIF approaches appear to be well aligned with the private sector's ambition of finding a path to fusion that is simpler and cheaper. This contrasts with the well-established large tokamak approach favoured by governments and as exemplified by ITER. Accordingly, following the surge of private money in pursuit of simpler cheaper fusion, such concepts have seen a resurgence of scientific interest, as well. In many individual cases, but certainly collectively, these two factors have combined to spur the return of focused public support for MIF. The ARPA-E ALPHA (Accelerating Low-cost Plasma Heating and Assembly) programme in the US, for example, supported around a dozen fusion developers and its success has spawned additional initiatives in 2020 that provide further support to existing, successful developers, as well as new entrants. Now, private investors, buttressed by funding and support from ARPA-E has created a unique innovation ecosystem in which entrepreneurs and scientists can now build, test and advance their concepts. Programmes such as those run by ARPA-E do not weaken MCF and ICF approaches. Reactor designs from all conceptual spaces—MCF, ICF and MIF—are encouraged, so long as they show the potential for improving the overall prospect of commercial fusion. The hope of such support is to unveil whether potentially faster, simpler, and cheaper routes to fusion exist (compared with incumbent approaches, particularly large tokamaks based on ITER)[1]. The consequence of all this is that the private fusion industry is now gaining momentum and is attracting talented scientists and ambitious investors alike to join the mission, which may yet become a race.

This chapter provides an overview of the alternative fusion approaches currently under development by private companies, by providing a brief profile outlining the mission for each. We note specifically how recently many companies have emerged, and these companies reflect the diversity, and the complexity, of fusion technological concepts under investigation. The list of such companies cannot be wholly inclusive, nor exhaustive, not least because several companies have launched during the short time in which this chapter was written. Furthermore, we choose to focus on companies dedicated to producing commercial energy services, for example as electricity or as heat for industrial processes[2]. The overall intention is to provide the reader with a sense of the diversity of efforts, and a view of the status of each company in their mission towards commercial fusion. Before we offer detail on these private companies, it is appropriate to ascertain the broader context of how and why a private fusion industry is emerging. A prior publication by the authors in [1] analyzed the underpinnings of this fusion 'paradigm shift' from an innovation perspective. This chapter instead focuses on the investment perspective.

7.2 Private investment in fusion

As noted in chapter 1, in May 2019, the emergence of small private sector fusion companies around the world attracted the attention of an edition of the British news magazine *The Economist* [2]. Perhaps more than any other step, this indicated that fusion innovation had crossed over from being a matter for the laboratory to something of potential interest in the board room. But why now are investors waking up to the commercial opportunities of nuclear fusion technology?

7.2.1 Private investment into clean energy technology

The energy sector—in particular, clean energy—is typically considered as too uncertain and high risk for venture capital [3]. This perception is for a variety of reasons, including that investors have to consider stringent tax policies, regulation, and geopolitical factors in the development of new energy technologies [4, 5]. In addition, investors are given cause to pause by the extremely large capital required. Investors note that fusion energy commercialisation will not provide a return on their investment on any near-term, or even medium-term, time horizon. In the US, the average proportion of total capital investment into the energy industry (including utilities) from 2015 to 2020 was just over 1%, compared with 30% and 7% for the software and biotechnology industries, respectively[3]. As a consequence of this lack

[1] Not in a dissimilar way to the NASA COTS programme, see [6].

[2] We note that some companies are interested in neutron or x-ray production, but such pathways to commercialisation are somewhat different. Such pathways are considered briefly in chapters 11 and 12.

[3] Figures calculated from the authors' analysis of data from PricewaterhouseCooper/National Venture Capital Association *MoneyTreeTM* Reports from 2015 to 2020 (https://www.pwc.com/us/en/industries/technology/moneytree.html).

of investor activity, much investment into energy technologies, particularly those at an early stage, is provided by governments. Typically, such government-funded programmes must support the technology until it is almost ready for commercial deployment. At this point, having passed over the so-called innovation 'valley of death' [4], in which technologies that are technically advanced but not yet viable for commercial development, the technology is then taken up by industry. This innovation model is known as the extended pipeline model, see [1, 4]. Private investors and organizations, by this time, can see returns on a time horizon that they are comfortable with and a technology that is developed enough to be perceived as lower risk. Gaddy *et al* [7] suggest that compared with software start-ups (see previously), clean energy technology investments:

> typically require a longer period of investment than the average software start-up. Venture capital firms expect their investments to mature on a 10-year time scale. The first five years involve stages of progressive investment and helping the projects scale, while the following five years are expected to return profits to the venture capital firm. Most endeavours in clean energy take 10 to 15 years for the technology to mature, at the end of which there is no guarantee that the technology will be profitable or even have a market.

This arrangement reflects the existing public–private energy sector innovation paradigm [4, 5]. Substantial investment is therefore needed to develop such transformative technologies. In private energy companies, however, a dichotomy exists. Shareholders are aware of the societal need for clean energy, but they are presented with the opportunities coming from large, established and profitable incumbents. All too often they are not willing to invest their discretionary resources into new technology when, in business terms, they are content with the incumbent one [4]. In its current form, therefore, the clean energy innovation paradigm is not working well.

7.2.2 Private investment in fusion: same but different

Noting that private fusion companies are pursuing a technology that is unlikely, even according to some of its most ardent supporters, to be commercialised for at least a decade; why has there been such a surge in support and investment? The view of Randall Volberg, a fellow at the American Security Project and an expert on fusion energy policy, is that [8]:

> the pool of venture capitalists with the appetite for frontier investments in fusion is very limited given the standard investment thesis is geared to realize returns within five years and for projects with less perceived technological risk.

Volberg points out that only a select few investors are keen to invest in fusion and, indeed, funding may yet dry up. Conversely, however, one voice from the fusion investment community, Roy Rothrock (formerly of Venrock), suggests that investors are interested in fusion and go against the grain of the traditional clean energy sector paradigm [9]:

> Why would a venture capitalist pursue a speculative line of technology? Governments have already poured tens of billions into fusion research for decades, with scientific progress, but little to show for it commercially. Aren't we [Venture Capitalists] supposed to be ruthlessly focused on finding ideas that we think can be brought to market with reasonable investments and in reasonable periods of time? Of course. But sometimes an opportunity presents itself that, if it works, can change the future forever and for everyone. Even if it is a long shot, it is worth pursuing.

Also important to mention is that private fusion enterprises are sitting on decades of publicly funded fusion research which, from a technical perspective, has been largely successful. This is even the case for some of the more novel MIF concepts under development, albeit to a much lesser extent than for the more heavily researched tokamaks. There are no cases, however, in which a fusion reactor design has been progressed towards a viable commercial product. According to the extended pipeline model of innovation which, as noted, very much prevails in the energy sector, investors would typically wait for a commercial prototype to exist, and to have been tested, before investing. Despite this reality, when it comes to fusion, it seems that investors are nevertheless keen to invest even at such an early stage of development. Why is this happening? Most obviously, perhaps it is because the size of the prize—fusion offers the potential of providing a carbon-free technology able to take a strong position in an energy market worth trillions of dollars. Accordingly, private fusion is upending the traditional linear model upon which the fusion programme has historically been developed, and arguably upon which all energy technologies have typically been developed [1]. The American Security Project captures the current zeitgeist with these words [10]:

> Up to now, fusion has predominantly been the province of large, government-funded research programmes, with most research funded through Universities or National Labs. Today, however, venture capital funds have begun investing in new start-up companies that are trying alternative approaches to fusion. Private companies are streamlining the pathway to a fusion power plant.

While currently innovation is being accelerated by the private sector—notwithstanding the continuing efforts of major national laboratories to the same end (see chapter 8)—as noted previously, it is clear that all commercialisation efforts build upon public sector research that came before. While the public sector research has been essential, some investors in the emerging fusion companies may see a public sector history of slow progress and disproportionate focus on the tokamak. Fusion investors also point to a change in attitude vis-à-vis investor appetite for risk into challenges that have long time horizons. They are becoming more open to making large investments—if they hold the prospect of huge economic and social payoffs[4]. As evidence, Amazon's Jeff Bezos, as well as Breakthrough Energy Ventures—a collection of high-net-worth investors interested in clean energy, including household names such as Bill Gates, Richard Branson, George Soros and Jack Ma—are investing in private fusion companies. It is evident that investors see fusion as a key technology that can address a pressing global challenge: the reduction of Green-House Gas (GHG) emissions and clean energy generation to prevent the potentially catastrophic impact of climate change.

Moreover, with the support of initiatives like ARPA-E that provides funding for several companies at once, investors are backing several companies in a 'multiple shots on target' approach[5]. Compared to ITER, which arguably places 'all the eggs in one basket' [11], private investors are spreading the risk into a portfolio of different conceptual approaches—all of which could succeed, or all of which could fail—in the hope that the overall chance of success is increased. In this vein, it was reported in 2018 by Dan Brunner of MIT that the mission of 'Fusion One Corporation', an early start-up focused on the Polywell (an inertial electrostatic confinement fusion reactor concept), was cancelled after modelling and experimentation proved that Q_{fus} with D-T could only achieve a maximum of 3.5[6] [12]. However, despite the fact that the original mission was a failure, the report argues that this exploration into the Polywell should be regarded as a success of the broader private fusion approach to identify the highest risks and work to retire them (i.e. prove they are impossible) at the lowest cost and in the quickest time. For more on the Polywell see chapter 11. We suggest that for a technology-oriented group within the investment community, there is an emerging and genuine belief that with the right financial support, viable commercial fusion can be achieved.

Beneficial synergies between the public sector and the emerging private sector continue (see chapters 8 and 9). However, it is important to remember that the individual companies and government laboratories are not in commercial competition with one another. In principle, government laboratories should be keen to see private sector success. However, there is a residual concern that a prominent private sector failure might somehow taint the fusion research mission as a whole and, in doing so, jeopardize future public funding. The private sector approach to innovation approach involves a willingness to take higher risks together with a focus on adopting rapid developments; concepts that are not common in publicly funded programmes, in which taxpayer dollars are being spent. While some aspects of the innovation approach that may yield much success for private fusion companies, particularly the desire to get to a commercial product as soon as possible, specific characteristics are

[4] This is, perhaps, not dissimilar to the way in which the Bill and Melinda Gates Foundation is investing large sums of capital to support development of a Malaria vaccine.

[5] With thanks to Eric Ingersoll, of Lucid Catalyst, for this analogy.

[6] For interest, Brunner states that this maximum of Q_{fus} of 3.5 applies to all electrostatic concepts.

seemingly at odds with the approach of science carried out by way of fusion programmes at government laboratories [1]. Therein, the focus is on research and development, to gain a fuller understanding of the science. It is not to take risks and proceed with haste. As such, the driving concerns of private fusion companies are fundamentally different from the public programme approach with its emphasis on failure avoidance and a culture of testing everything before trying anything[7]. Despite this, there is a need for improved synergy between the public and private sector, to allay fears of mutual failure and to boost the chance of mutual success.

To this end, we suggest that an 'innovation orchards' approach [3], in which private companies work with publicly funded laboratories, universities and also established industry (for example, large engineering companies) to leverage the best ideas, expertize and capabilities from each. Using this model as a basis for collaboration could result in the development of new and better technologies[8]. Indeed, this approach is inherent in many fusion companies already who, having spun out of universities or laboratories, can lean on their former parent organizations to plug gaps in their knowledge, skills, and capabilities. Others are able to strike up partnerships with universities and laboratories that have conducted research relevant to their specific fusion concept, and who may thus be keen to support the expanded development of their ideas and technical knowledge in the commercial environment provided by a private company. Ergo, while investment for fusion companies from venture capital is fundamental for success, part of successful fusion innovation can be the readiness and willingness to leverage ideas from others inside the fusion ecosystem.

7.3 Private fusion companies

Details of 18 private fusion companies from around the world are provided in table 7.1 (listed in order of date founded). Whilst the list is comprehensive, it is not exhaustive[9].

It is interesting to note the diversity of fusion approaches being explored, and also the disparity in investment between companies. Some are yet to receive significant backing because they have only recently launched, and others have only been awarded relatively small quantities of funding to demonstrate initial proof of principle for their concepts. In the latter case, companies are funded only by seed capital or, more often than not, by specific US ARPA-E programmes[10]. A handful of the companies, including some that are pursuing ostensibly wholly novel reactor concepts (i.e. non-tokamak approaches), have raised hundreds of millions of dollars, and have been established for more than a decade. A notable exception to this is

[7] Government laboratories can be more risk taking than this characterization might suggest. The colourful story of the START tokamak, as operated at the UK's Culham Laboratory in the 1990s, provides perhaps the most apt example, as a key part of the START narrative is that the device was largely developed from surplus components and put together by a team of enthusiastic scientists during their free time.

[8] The *Innovation Orchards* model is not dissimilar to the *Triple Helix model* of innovation, see [45].

[9] For those seeking a complete list of private fusion companies we recommend: www.fusionenergybase.com Readers may also refer to the interactive maps in [13, 14], which are regularly updated.

[10] ARPA-E has three programmes focused on advanced fusion development: ALPHA, BETHE and GAMOW. For further information, refer to https://arpa-e.energy.gov/.

Table 7.1. Private fusion companies (location, launch date, concept, and investment). Companies listed in order of date founded. Investment data as reported up to September 2020. Data is limited to available information from publicly visible sources only. All monetary values are in US Dollars ($).

Company	Location	Date founded	Concept	Investment (estimated as of August 2020)
TAE Technologies (formerly Tri-Alpha Energy)	Orange County, CA, United States	1998	Field Reversed Configuration (FRC)	~700 M
General Fusion	Vancouver, BC, Canada	2002	Magnetically confined acoustic compression	~200 M
Lawrenceville Plasma Physics	Middlesex, NJ, United States	2003	Dense Plasma Focus (DPF)	~7 M
MIFTI Fusion	Tustin, CA, United States	2008	Z-Pinch	~8 M
Tokamak Energy Ltd	Oxford, United Kingdom	2009	Spherical Tokamak (with HTS magnets)	~150 M
First Light Fusion Ltd	Oxford, United Kingdom	2011	Hypervelocity Inertial Confinement	~73 M
Proton Scientific	Oak Ridge, TN, United States	2012	Electron Beam Fusion (EBF)	~5 M
Helion Energy	Redmond, WA, United States	2013	Field Reversed Configuration (FRC)	~12 M
Lockheed Martin Skunk Works	Palmdale, CA, United States	~2013	Magnetic Cusp Device	Undisclosed[a]
CTFusion	Seattle, WA, United States	2015	Spheromak	3 M
Agni Fusion	Olympia, WA, United States	2017	Beam-Target Fusion (MIF)	Undisclosed
Commonwealth Fusion Systems	Cambridge, MA, United States	2017	Tokamak (with HTS magnets)	~200 M
Compact Fusion Systems	Santa Fe, NM, United States	2017	Stabilized Linear Compression	~4.4 M
Hyperjet Fusion	Chantilly, VA, United States	2017	Plasma Jet Liner Compression	~2 M
ZaP Energy Inc.	Seattle, WA, United States	2018	Sheared-Flow-Stabilized Z-pinch	~20 M
HB11 Energy	Freshwater, New South Wales, Australia	2019	Chirped Pulse Amplification (CPA) p-B11	Undisclosed
Marvel Fusion	Munich, Bavaria	2019	Chirped Pulse Amplification (CPA)	Undisclosed
Renaissance Fusion	Grenoble, Auvergne-Rhone-Alpes, France	2019	Stellarator	Undisclosed

[a]Lockheed Martin's Skunk Works fusion project is being funded with internal investment from the company

Commonwealth Fusion Systems which, which despite only launching in late 2017, is now reported to have raised more than \$200 M[11].

The next parts of this chapter will provide an overview of the private fusion companies who are the recipients of the surge of private investment that was discussed in the previous section. The following subsections detail each company's reactor concept, mission, progress and vision for commercialisation listed in order of date founded (per table 7.1).

Note that entries for Tokamak Energy and for CANDY GPI-Hamamatsu are not included in this chapter, as details of their efforts in MCF and ICF respectively have been provided in chapters 5 and 6.

7.3.1 TAE Technologies

TAE Technologies, formerly known as Tri-Alpha Energy, was founded in 2001 in California, USA. The company is developing a fusion reactor concept based on the field reversed configuration (FRC) that will operate on the p-B11 fuel cycle. The p-B11 (hydrogen- or proton-boron) fusion reaction—the fusion of hydrogen (H) with the boron isotope ^{11}B—is what is known as an aneutronic fusion reaction. This means that it does not produce neutrons and, therefore, produces minimal radioactive waste (see chapter 10). The reaction can also generate electricity directly, as the reaction produces three charged alpha particles, meaning there is no need for a thermal power cycle. Unlike tritium, which is required for D-T fusion, both hydrogen and boron are readily available and not radioactive. As such, it is p-B11 fusion that is seen by some as the true 'holy grail' when it comes to fusion energy. That said, while D-T fusion is the 'easiest' reaction to achieve (in terms of the conditions required for fusion to occur), p-B11 fusion is probably the most difficult. However, proponents think that if the physics challenges can be overcome, and a p-B11 reactor is proven possible, it may present the best option for commercial fusion energy.

TAE's reactor configuration sees beams of high-energy hydrogen atoms driven towards one another to produce a self-contained plasma in a reactor chamber surrounded by strong magnetic coils. TAE's FRC uses closed magnetic fields rather than a physical barrier, as in a tokamak (see chapter 2), to contain the plasma. The magnets work to stabilize the plasma that is produced upon the two beams merging [13]. As the p-B11 reaction requires extremely high temperatures, TAE's reactor concept also uses high-energy particle accelerators, which are placed tangentially to the plasma to both increase the energy to it, and to 'spin' and stabilise it, improving confinement. The TAE reactor configuration is shown in figure 7.1.

In 2015, TAE sustained a plasma for long enough for fusion to occur. In their current machine, 'Norman', operational since 2017 and named posthumously after the company's co-founder Norman Rostoker, TAE is now attempting to generate a plasma hot enough for fusion to occur. When the company has achieved this milestone, it will try to combine the two to produce a plasma that lasts 'long enough', and that is 'hot enough'.

[11] All monetary values are hereafter provided in US Dollars (\$).

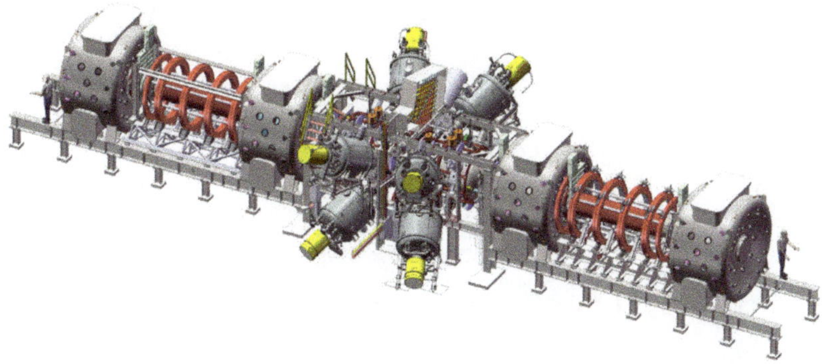

Figure 7.1. TAE Technologies' FRC reactor. The plasma producing beams are shown at each end of the diagram (left and right), and the chamber, surrounded by magnets, is shown in the centre. The perpendicular accelerators are shown in the centre of the diagram. Reproduced from [14] with the permission of AIP Publishing.

The goal for the company is to produce an economically competitive, baseload power plant that generates in the region of 200 to 500 MW of electricity. The company considers that the compact size, the use of the p-B11 aneutronic fuel cycle, and the direct conversion to electricity[12] makes their pathway to commercial fusion energy optimal compared with its competitors. The company has been supported by over $700 M of private investment since its launch over 20 years ago, and it has filed over 800 patents in that time. TAE is backed by private sector giants, Google, who contend that machine learning will play a key role in the next-steps of TAE's reactor technology development [15].

In 2018, the company also created a spin-off called 'TAE Life Sciences', which is focused on developing aspects of the TAE technology for non-energy purposes; specifically, for cancer treatment via Boron Neutron Capture Therapy (BNCT)[13].

7.3.2 General Fusion

General Fusion was founded by Michel Laberge in 2002, in British Colombia, Canada. The company is pursuing a route to fusion energy via a novel acoustically driven MIF reactor approach (General Fusion refers to its approach as 'Magnetised Target Fusion' (MTF) [16]). A self-confined magnetized spherical plasma containing deuterium and tritium fuel is injected into a cavity at the centre of a chamber surrounded by a liner made of liquid lead-lithium metal. Rapidly following plasma injection, an array of acoustically driven pistons surrounding the plasma produces an acoustic pressure wave which accelerates the liner towards the plasma to instantaneously compress and heat it [17]. General Fusion suggests that the plasma's volume decreases by three orders of magnitude, and correspondingly the density, temperature, and magnetic field are increased to the point at which fusion occurs. The

[12] The original name of the company, 'Tri-Alpha Energy', is a nod to the potential for direct electricity conversion from p-B11 fusion, which yields three alpha particles.

[13] BNCT is a method by which boron, placed inside the human body to accumulate in cancerous cells, is targeted by neutrons to destroy the cancer cells by way of ultra-localized radiation.

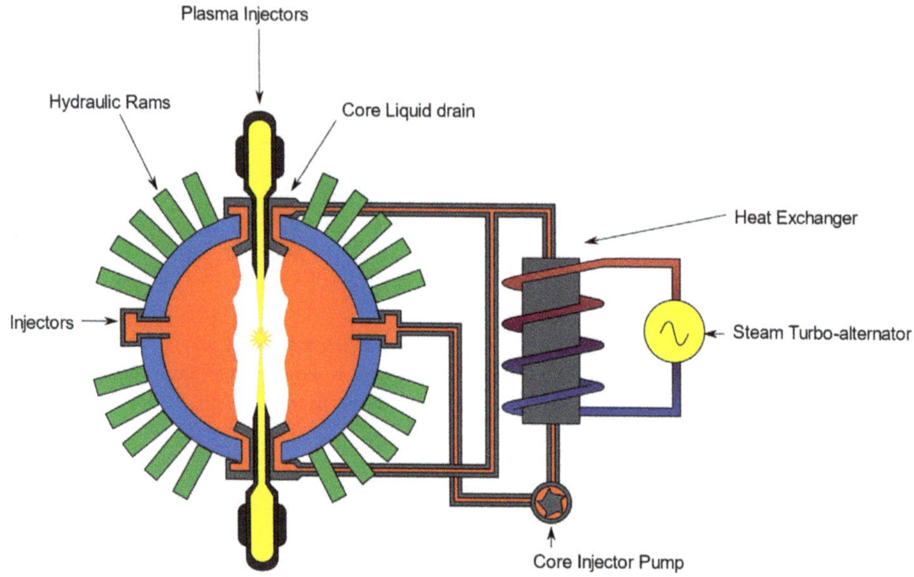

Figure 7.2. General Fusion's fusion reactor concept, showing plasma injectors, acoustic drivers (denoted as 'hydraulic rams'), and the injector for the metal liner, connected to the power generation systems (including heat exchanger). Image created by Evan Mason and reproduced under CC BY-SA 3.0 license (Wikimedia Commons).

neutrons from the fusion reaction carry their energy to the liquid metal lithium-lead liner, which transfers it (as heat) via a heat exchanger to produce electricity. The metal liner, containing lithium, has the dual function—as with all other D-T fusion reactor designs—of producing tritium. These stages are repeated such that the reactor operates cyclically, in the same way as an ICF reactor, but with lower frequency (see chapter 2). The General Fusion reactor configuration is shown in figure 7.2.

Alongside numerical simulations of General Fusion's approach, the company is currently focused on parallel R&D on its component technologies that make up the core reactor systems (the driver, liner and target—see chapter 2). Separate projects to develop the acoustic drivers, plasma injectors, and liquid lead-lithium pumping technology will later be integrated into a prototype reactor. The company believes its approach presents the most efficient and fastest path to fusion, and is focused on low-cost electricity generation as its primary commercial goal. The company has raised around $200 M and is backed by investors such as Jeff Bezos, of Amazon.

7.3.3 Lawrenceville Plasma Physics

Lawrenceville Plasma Physics (LPP), based in New Jersey, USA, was founded by Eric Lerner, a plasma scientist who also has an interest in plasma cosmology, including on theories of the origins of the Universe, including an alternative theory to The Big Bang. LPP hopes to achieve commercial fusion success via a Dense Plasma Focus (DPF) reactor which, like TAE technologies, will use an aneutronic hydrogen-boron fuel cycle [18]. The energy from the DPF will be produced via a beam of charged particles (helium nuclei) which can be converted directly into electricity. While the DPF has been studied as an experimental reactor since the 1960s, LPP is the only major initiative pursuing this particular variant of the aneutronic hydrogen-boron approach. The DPF approach, though comprising a simple set up, is still scientifically and technically challenging. The following explanation is from the authors' interpretation of information from the LPP Fusion website and the publication by Lerner *et al* in [18], see text box.

> The DPF exploits a series of naturally occurring plasma instabilities to funnel electric current into a magnetic pinch. The DPF device consists of two cylindrical metal electrodes nested inside each other. The electrodes are enclosed in a vacuum chamber with a low-pressure gas filling the space between them. A pulse of electricity from a capacitor bank (an energy storage device) is discharged across the electrodes. For a few millionths of a second, intense current flows from the outer to the inner electrode via the low-pressure gas, which heats it to create a magnetic field. The current in the plasma then forms into a sheath of tiny filaments—vortices of hot, electrically-conducting plasma—which is guided by the magnetic field. The sheath travels to the end of the inner electrode where the magnetic fields produced by the currents pinch and twist the plasma into a tiny, dense plasmoid that is only a few thousandths of an inch in size. The magnetic fields very quickly collapse to induce an electric field which causes a beam of electrons to flow in one direction and a beam of ions in the other. The electron beam heats the plasmoid to extremely high temperatures, and the collisions of the electrons with the ions generate a pulse of highly intense X-rays. Energy is transferred from the electrons to the ions (under the magnetic field effect) along the high-density plasmoid with sufficient energy to yield a fusion reaction. The result is a high-power ion beam (where the power from the ion beam is higher than the electrical current required to initiate it). The beam of charged particles is directed into a decelerator which acts like a particle accelerator in reverse. In other words, instead of using electricity to accelerate charged particles, it decelerates the charged particles to generate electricity (some of which is recycled to power the next fusion pulse). According to LPP some of the X-ray energy produced by the plasmoid can also be directly converted to electricity through the photoelectric effect (as in solar panels)[14].

LPP has produced several experimental reactors, including most recently the FF-2B device that the company hopes will achieve net fusion power gain in the coming years [19]. Subject to experimental success and next-stage investment, the company hopes to develop a first commercial DPF prototype that will produce 5 MW net electricity. While LPP is targeting electricity generation, the company is also exploring the possibility of its generators for use in large-scale transportation. An alternative commercialisation route is for the DPF to act as a portable, low-cost intense x-ray source. Given that parameters of the DPF could be altered to maximize x-ray output, the technology could be used for non-destructive inspection of materials, including for civil engineering and infrastructure (bridges, roads, buildings etc). LPP suggest that the DPF for x-ray applications is ready for mass

[14] For a more detailed description of the DPF approach, see [18].

production and that this aspect of the technology is for sale—the company cites that this approach is analogous to the Wright Brothers, who ran a bicycle shop as a means to fund their flying machine experiments! The company is currently backed by over 500 private investors and is the only fusion start-up, as far as the authors are aware, that has employed crowdfunding as an investment mechanism. LPP also received funding from NASA's Jet Propulsion Laboratory in the 1990s.

7.3.4 MIFTI Fusion

Following several decades of research into the Z-Pinch at the University of California, Irvine, CA, US, Magneto-Inertial Fusion Technologies, Inc. (MIFTI) was established in the early 2000s to develop the 'Staged Z-pinch' fusion reactor design. The company was founded by Hafiz Rahman, a leading fusion research scientist at University of California, Irvine, along with Gerald Simmons and Mohammad Arshad, two businessmen, in 2008. MIFTI's concept is a variation of the Z-pinch—a type of fusion reactor studied in-depth in programmes across the world at the dawn of fusion research in the 1950s—that overcomes instabilities to yield a stable and longer-lived plasma. The Staged Z-Pinch works by simultaneously heating and compressing a fuel target to produce a fusion reaction. The fuel target is comprised of deuterium and tritium enclosed within a high-Z liner, typically made from a heavy noble gas such as argon or krypton. The fuel target is then subjected to an intense, pulsed electrical current for a fraction of a millisecond. This current generates a magnetic field that rapidly compresses the gas within the target, which collapses the liner inwards to form a stable plasma. It is with this approach that MIFTI claims to avoid the aforementioned instabilities that plagued original research into the Z-pinch. After initial ohmic heating, supersonic compression heats the fuel to very high temperatures such that fusion conditions are achieved. Once fusion occurs, the reaction propagates—in fact, in a way not dissimilar to ignition in an ICF reactor—and the particles deposit their energy inside the plasma to sustain a D-T fusion reaction with a large power gain over just a millionth of a second.

MIFTI point to detailed computer simulations and experimental observations, suggesting that the Staged Z-pinch can produce the most stable fusion reaction of all concepts under development today. Although like other fusion companies the long-term objective is electricity generation, MIFTI is targeting different, near-term 'lower risk' commercialisation pathways for the staged Z-pinch. In particular, the company have identified medical isotope production—an industry worth over \$5 B, globally—as well as the transmutation of long-lived nuclear waste as promising avenues. Research when the concept was under development at the University of California, Irvine was supported by the US Air Force. Since spinning out of the university, the company has received an ARPA-E grant (2015) and has also been supported by Strong Atomics LLC, a group of venture capitalists from Silicon Valley who are interested in commercial fusion.

7.3.5 First Light Fusion Ltd

First Light Fusion Ltd, based near Oxford, UK, was spun out of the University of Oxford in 2011 by Nicholas Hawker and Yiannis Ventikos. The company builds upon the founder's research into the hydrodynamics of shock-driven cavity collapse. The company is exploring a reactor concept that is based on the same fundamental physics that underpins ICF but is taking a significantly different approach to developing a reactor. The company contends that most ICF reactors attempt to achieve cylindrical implosion of a fuel target to provide uniform and efficient compression but struggle to deal with the instabilities involved; in particular, the Rayleigh–Taylor instability (as mentioned in chapter 2). Instead of avoiding these instabilities, First Light's approach seeks to harness them. The company's concept works via using an intense shock wave to crush a uniquely designed gas-filled cavity which undergoes an asymmetric collapse —using a mechanism similar to the pistol shrimp claw (see [20])—to achieve an implosion that yields temperatures and compression inside the cavity high enough for inertial fusion to occur. First Light aims to create such conditions by firing a projectile from a two-stage gas gun at the gas-filled cavity [21], which is shaped to achieve maximum and uniform compression for fusion [20].

The company is hoping to demonstrate fusion conditions in its next test machine, called 'Machine 3', and suggests that its modelling and experimental testing towards this goal is thus far consistent. First Light expects it to be simpler to engineer their concept into a commercial fusion reactor than tokamak-type concepts. In particular First Light stresses that their concept does not require the powerful magnets or lasers associated with many other fusion concepts under development. This, the company suggests, will mean more cost-effective electricity generation. Work to develop a conceptual power plant design to achieve this goal is currently underway [20]. The company has raised over $70 M, backed by IP Group PLC and Parkwalk Advisors, investment firms focused on supporting high-tech endeavours.

7.3.6 Proton Scientific Inc.

Proton Scientific Inc. was established in Champaign, IL, USA in 2012 to develop a pulsed power electron beam generator, an ICF concept. The company has more recently relocated to Oak Ridge, TN, US The company, using its IVR3 experimental electron beam generator—codenamed 'Thunderbird'—will produce a high-energy electron beam that will be focused onto a microscopic target, in a similar way to fast ignition concepts (refer to the description in chapter 2). The beam will strike the solid target with an extremely large power density, in the region of 10 TW mm^{-2}, to achieve the conditions for inertial fusion. Proton Scientific believes that the electron beam produced by Thunderbird can provide power densities that were previously only thought possible via large laser pulses. The company also suggests that the Thunderbird beam can deposit its energy to the target to achieve a uniform compression and burn. The company, therefore, intends to reduce the cost and complexity of ICF drivers in its approach.

Proton Scientific intends to develop a full electron beam fusion prototype in the early 2020s [22]. The company has not disclosed the levels or sources of its investment.

7.3.7 Helion Energy

Helion Energy was founded in Redmond, WA, USA, in 2013, with a Field-Reverse Configuration (FRC) concept that the company calls the 'Fusion Engine'. The Fusion Engine concept combines the stability of magnetic fusion with the high temperatures of inertial fusion to produce a pulsed system. The company contends this combination allows for a more compact reactor and a more cost-effective route to fusion energy. The concept, being an FRC, is not dissimilar to the TAE Technologies approach. Helion's Fusion Engine functions by creating two FRC plasmoids which are accelerated to extremely high velocities—in the order of 300 km s^{-1}—towards each other [23]. The two plasmoids collide and merge at supersonic speeds, in the process converting kinetic into thermal energy. At the point of merging, the plasmoids are compressed by strong magnetic fields, to create the combined conditions of high pressure and temperatures required for fusion. Helion suggests that the conditions expected to be achieved by the fusion engine are enough to allow a D–D or even D–^{3}He fusion reaction. Whilst D–D fusion reactions produce neutrons and tritium, like p–B11 fusion, D–^{3}He fusion is an aneutronic fusion reaction. Helion is thus attempting to produce a fusion reactor that does not produce neutrons, and which does not require tritium, and hence, in turn, does not require tritium breeding (tritium breeding is explained in detail in chapter 10).

The company hopes to realize a route to fusion that is thus simpler and cheaper than D-T approaches, including to generate electricity by direct conversion; similar to both TAE Technologies and LPP Fusion. Helion has raised around $12 M from seed investors, including Y-combinator, a US-based seed funding technology accelerator, and the company was also awarded ARPA-E funding under the ALPHA programme in 2015 [23].

7.3.8 Lockheed Martin's Skunk Works

Lockheed's Skunk Works division is known for developing revolutionary technologies such as the SR-71 Blackbird spy-plane. The culture of Skunk Works is special, being characterized over many decades by an abundance of autonomy, extremely limited bureaucracy and intense secrecy. Arguably, the Skunk Works approach has provided a blueprint for innovation that has been emulated for the development of advanced technologies beyond the defence industry, including in Silicon Valley. As such, in 2013, when the company announced that it had commissioned its Skunk Works division to embark on a fusion reactor development programme, it sent a shock wave through the fusion world. While military involvement in fusion R&D has taken place in many forms in previous decades, most notably for ICF, that a privately held defence company was pursuing the development of a novel and compact (i.e. small) fusion reactor was something extraordinary. Noting that it is very difficult to say anything insightful about such secretive programmes, it is certainly the case that the public understanding of the relationship between fusion science and technology and military ambitions has largely revolved around the issues of thermonuclear weapons. This interest from Lockheed Martin is potentially military, but it is seemingly not connected to the H-bomb in any way. It is more likely to be driven by a desire to achieve a powerful, compact, and reliable energy source for challenging environments. Ostensibly, Lockheed Martin stands out from the crowd amongst private fusion companies. Investment is likely in the tens of millions of dollars, but the amount will most certainly remain unknown. More importantly, however, it is certainly an outlier in terms of its development approach, given that the reactor is being developed under its famously secretive programme. As can be expected, therefore, there is limited information on Lockheed Martin's reactor concept in the public domain.

The fusion project at Lockheed is headed by Thomas McGuire. The team is developing an entirely novel fusion reactor based on a 'high beta' concept. At the risk of slight over-simplification, beta is the ratio of plasma pressure to the magnetic pressure, and the concept employs a high fraction of the magnetic field pressure which enables the reactor to be compact. For this reason, the company contends this will produce efficient and stable plasmas, and it is why Lockheed has named it the 'Compact Fusion Reactor' (CFR) [24]. The CFR concept uses sets of ring mirrors and magnets that produce a type of magnetic field known as a diamagnetic cusp. The magnetic forces rapidly change direction and push the atomic nuclei in the plasma towards the centre of a cylindrical vessel. External magnets then push the nuclei back towards the ends of the vessel. The plasma pressure causes the plasma to expand and the field becomes stronger at the plasma edge to increase plasma containment. Lockheed suggests that the concept has no net current, which means that plasma instabilities are avoided. The dense target plasma produced in the CFR will be heated via neutral beam injection (see chapter 2 and [25]). The configuration of the reactor is shown in figure 7.3.

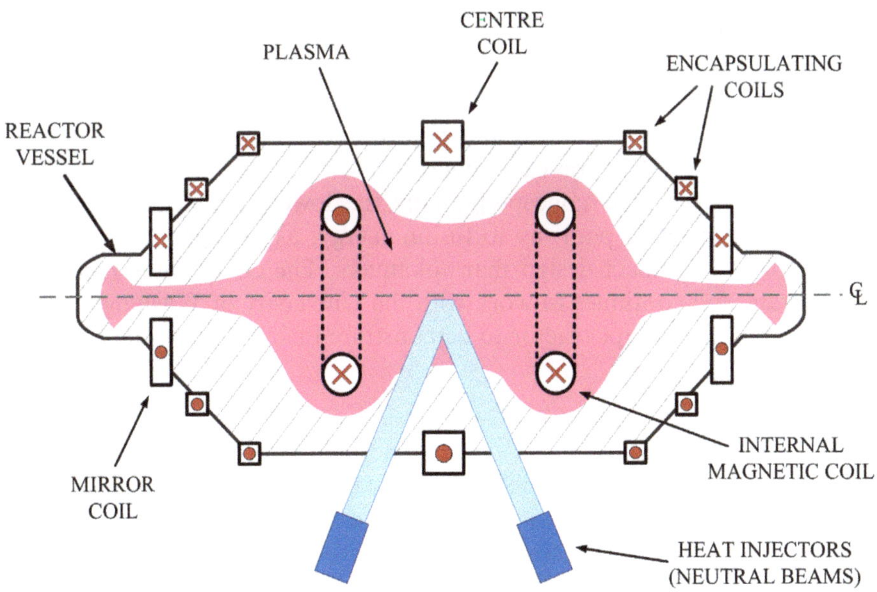

Figure 7.3. Cross-section view of the Lockheed Martin Skunk Works CFR configuration. Note that the current in the internal coils is polarised in the opposite direction to all the external coils. Diagram is author's interpretation of the reactor, based on a patent by Lockheed Martin [26] and personal communication with Thomas McGuire, of Lockheed Martin. The authors also thank Dr McGuire for feedback and clarification on the design and function of the CFR.

Lockheed has developed, built, and tested several reactors, and has published details on the results of the T-4 and T-4B reactor experiments. It is now developing a bigger and more powerful test reactor called T5. Regardless of the technical feasibility of the concept, the company has demonstrated that it can develop on rapid cycles, which the company says is an important tool to 'build up momentum' in their development approach [27]. We have previously suggested that an agile approach and compact reactor size are key factors in accelerating the development of fusion [1].

Lockheed suggests that the heat from their reactor will drive turbine generators to produce electricity. In addition, the company indicates that it could be used for desalination or propulsion, with applications in aerospace, marine or even space travel [28]. Given that Lockheed is heavily involved in military technology, this is rather unsurprising. As before, while no details are available on the level of financial support being given to the fusion programme in terms of funding or personnel, reflecting the nature of the Skunk Works division, the project appears to receive continued funding and backing from the company.

7.3.9 CTFusion

Decades of research at the University of Washington, WA, USA, has been focused on the spheromak, a type of MCF reactor that sustains stable plasmas by concurrently driving a plasma current near the plasma chamber wall to induce a magnetic field. While spheromaks and tokamaks (including spherical tokamaks) involve similar technologies, they are, in fact, altogether different concepts. In 2015, CTFusion was spun out of the university as a private company to develop the reactor. Led by scientists Tom Jarboe and Derek Sutherland, the company has developed a new type of current drive called Imposed-Dynamo Current Drive (IDCD). The company have heralded the IDCD as a breakthrough technology that enables steady-state magnetic fusion operation by avoiding plasma instabilities [29]. The technology turns the conventional spheromak into what CTFusion call a dynomak.

CTFusion believes their concept presents the lowest technical risk and most economically competitive pathway to fusion energy, as the spheromak provides a simpler and more compact design than tokamaks. The company is funded by an ARPA-E grant and continues to work with the University of Washington towards building a proof-of-concept reactor to demonstrate fusion power gain [30].

7.3.10 Agni Fusion

Agni Fusion is a start-up based in Washington State, USA. Agni's reactor was first developed as part of a high school science project, in which co-founder Dmitri Hopkins presented an earlier version of the concept to US President Barack Obama after winning an international science fair. The Agni reactor is a linear configuration in which deuterium and helium-3 fuels are injected into a chamber where they are ionized. The ionized fuel mixture is passed through a magnetic focusing lens to create an ion beam, in which the density of the fuel mixture is increased to induce D-^{3}He fusion via ionic heating [31]. The result is a high-power and high-density ion beam. That ion beam is then targeted at a highly dense lithium borotritide target (a compound of lithium, boron, and tritium), striking it with enough kinetic energy to induce D–T fusion.

Agni intends to use its reactor to produce electricity, however, the company also indicates that an alternative, but additional, application is to use the reactor's fusion neutrons for nuclear waste remediation. The company has not provided details of any investors supporting its development and, as far as we are aware, has not published information on its progress.

7.3.11 Commonwealth Fusion Systems

Commonwealth Fusion Systems (CFS), based in Cambridge, MA, USA, was founded in 2017 as a spin-off from Massachusetts Institute of Technology (MIT). CFS, which is supported by MIT's Plasma Science and Fusion Center and its scientists, is pursuing a pathway to commercial fusion via a high-field conventional aspect ratio tokamak using HTS magnets. The high-field tokamak concept has strong foundations at MIT, with designs dating back as early as the 1970s. For example, see 'Ignitor', a compact tokamak reactor still under development by Bruno Coppi and Ron Parker [32]. MIT also operated Alcator C-Mod, a medium-sized tokamak, with much success[15]. CFS thus seeks to leverage the depth of knowledge and experience in operating tokamaks to accelerate commercial fusion.

The company's first step is to develop functional HTS magnets; the key enabling technology for their approach. Following successful demonstrations of its HTS magnets, CFS intends to build a conventional aspect ratio experimental tokamak equipped with those magnets to operate at high field. That tokamak, dubbed 'SPARC', currently under design, will demonstrate operation at high magnetic field and will also demonstrate net fusion power gain. A conceptual image of SPARC is shown in figure 7.4. The company aims to achieve its milestone of fusion power gain by 2025, but —like ITER, which has a similar goal that it hopes to achieve a decade later (2035)— SPARC will not produce net energy from fusion (see chapter 2). It will, however, provide a proof of principle that will pave the way for the next-step reactor, 'ARC' (standing for Affordable-Robust-Compact)[16], a full-scale prototype that will scale to a first commercial fusion power plant. The original concept for ARC was born as a result of a large collaborative student design project at MIT in the mid-2010s, led by Dennis Whyte [33]. Several of the students involved in that MIT project are now working with CFS.

The company is supported by investors including ENI, an Italian oil company, and Breakthrough Energy Ventures which is a philanthropic foundation supported by backers such as Bill Gates and Jack Ma. The company has raised over $200 M in

[15] Perhaps the most notable achievement of MIT's Alcator C-Mod is that it broke the record for highest plasma pressure in a tokamak in 2016, when it embarked upon its final experimental run before it was shut down due to funding cuts.

[16] We are amused to note the perhaps coincidental terminology from CFS of the 'palladium core' fusion reactor developed by the comic book superhero Iron Man (Tony Stark), which was also dubbed 'the ARC reactor', in the Marvel movie *Iron Man* (2008).

Figure 7.4. Artist's impression of SPARC being commissioned. Image created by Ken Filar, Research Associate at MIT, and reproduced under CC BY-SA 4.0 license (Wikimedia Commons).

the three years since it was founded, making it the fastest-growing private fusion company in the world.

7.3.12 Compact Fusion Systems

Based in Santa Fe, NM, USA, Compact Fusion Systems are exploring a concept called the stabilized liner compressor that was originally conceived over 50 years ago. The company was founded in 2017 by a team of fusion scientists and engineers —Simon Woodruff, Peter Turchi, and Ronald Miller—to develop the concept for commercial application. The stabilized liner compressor operates by injecting two plasmoids towards each other in the axial direction, in a similar way to that of Helion and TAE Technologies. The plasmoids merge in the centre of the reactor to form a single plasma. A rotating hollow cylinder of liquid metal surrounding the formed plasma is then rapidly imploded onto the plasma via an array of gas-driven pistons. As in all MIF approaches (see chapter 2), this causes the plasma to be instantaneously compressed and heated to conditions for fusion to occur. These stages are shown in figure 7.5. The company, partnering with several laboratories and academic institutions, has so far developed the technology through a series of ARPA-E grants and seed funding. The company is now seeking further private investment to support the engineering design of the plasma injection systems and the liner compressor, both of which will allow the build of their prototype device to demonstrate proof of concept [34].

7.3.13 Hyperjet Fusion

Hyperjet fusion corporation, based in Chantilly, VA, USA, was launched in 2017 as a spin-off from successful research at Los Alamos National Laboratory into Plasma Jet Driven Magneto Fusion (PJMIF). The approach was developed in support of the laboratory's Plasma Liner Experiment (PLX). The PJMIF approach pursued by

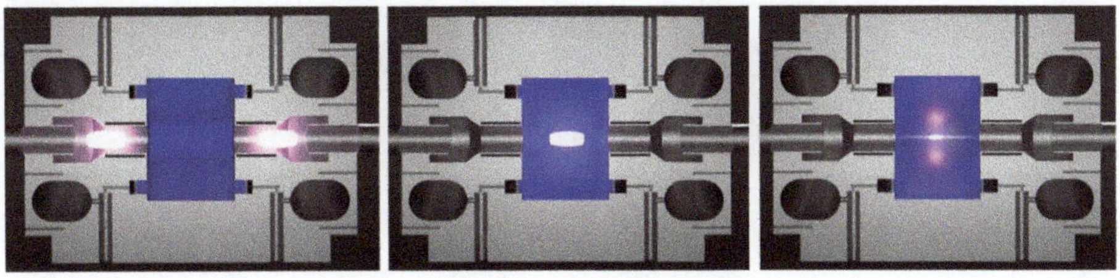

Figure 7.5. Stages to achieve liner compressed fusion in the Compact Fusion Systems concept (stabilized liner compressor). Image credit: Simon Woodruff, Compact Fusion Systems, used with permission.

Hyperjet employs a spherical array of coaxial plasma jet guns—around 600 of them—which merge to create a spherically imploding plasma liner that compresses and heats a magnetized target plasma [35]. The reactor configuration is shown in figure 7.6. Hyperjet believes that the benefits of a high shot repetition rate and fast implosion speed on the PJMIF, as well as the avoidance of instabilities and thermal losses, will open up a novel pathway to fusion power that allows rapid reactor deployment at low-cost.

The company appears to have benefitted from extensive research in the original publicly-funded PLX fusion programme and has consequently leveraged knowledge, expert personnel, facilities, and equipment from its time in development at Los Alamos. Several US government grants funded development at Los Alamos, including from ARPA-E [37]. However, the company itself has been provided with seed funding from Strong Atomics LLC in 2017 and again from ARPA-E more recently in 2020. Hyperjet foresees that its fusion reactor has the potential to provide baseload energy, but that the technology may have unique aerospace and industrial applications, too [38].

7.3.14 HB11 Energy

Although HB11 Energy, based in New South Wales, Australia, officially launched in 2019, research into the company's fusion approach has been a focus for the duration of the 60-year career of Heinrich Hora, founder of the company. HB11 span out of Hora's research programme at the University of New South Wales to pursue commercial development of its aneutronic hydrogen-boron (hence 'HB11') fusion with the use of Chirped Pulse Amplification (CPA) lasers, which are lasers that can provide petawatt power in very short pulses. CPA lasers were developed in the 1980s, but have undergone much development since that time. Most notably, they were the subject of the 2018 Nobel prize in physics. The HB11 approach uses a high-power CPA laser pulse directed at a specially designed fuel target made from hydrogen and boron. The laser first strikes one end of the fuel target, which is rich in hydrogen, to produce a hydrogen plasma that then accelerates towards the boron part of the fuel target. This plasma is confined by a magnetic field which is created by a second laser beam with a pulse duration of a few nanoseconds (ns), which causes magnetic trapping. Together, the two lasers create what the company refers to as an 'avalanche' of hydrogen–boron fusion reactions, which yield charged helium particles which can be used for direct electricity generation [39]. HB11 has published

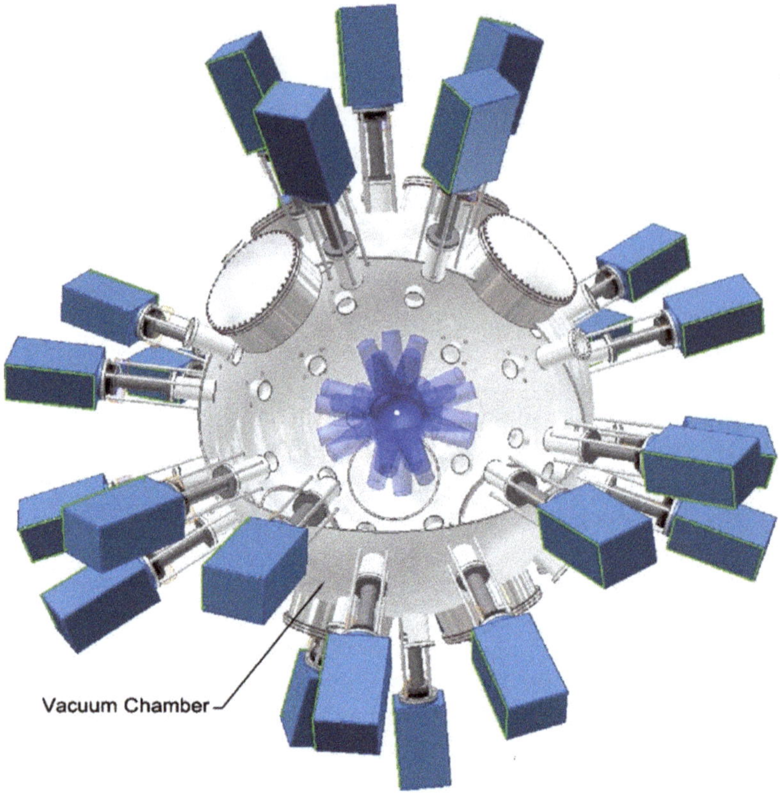

Figure 7.6. HyperJet's PLX fusion reactor concept. The box structures surrounding the vacuum chamber represent the capacitors which are attached to plasma guns, which fire a converging plasma liner towards a magnetized plasma target (shown in the centre). Reproduced from [36] with permission of AIP Publishing.

plans to build a prototype reactor, based on Hora's research—including experimental data, peer-reviewed publications, and patents. At the time of writing, no information is available regarding the company's investment status.

7.3.15 ZaP Energy

In 2018, scientists Uri Shumlak and Brian Nelson, along with entrepreneur-investor Benj Conway, launched Zap Energy as a spin-off to develop a novel flow Z-pinch fusion reactor that was being researched the University of Washington, WA, USA. As with MIFTI, a flow Z-pinch reactor is a fusion reactor with a linear configuration that uses a self-confining magnetic field—that is, with no applied external magnetic field—which is generated by the axial current to produce a high-temperature, high-density plasma. The concept has been extensively researched, but it was originally considered unsuitable as a fusion energy concept due to fusion energy concept due to 'sausage' and 'kink' instabilities that form in the plasma. However, an approach to use sheared axial flows, where the plasma flows at different velocities across the radius of the plasma, developed by the University of Washington and Lawrence

Livermore National Laboratory, and tested on the Fusion Z-pinch Experiment (FuZE), demonstrated long-lived plasmas and sustained neutron production [40]. Now at ZaP Energy, the flow Z-pinch concept is being developed towards commercial application. ZaP believe it will provide the simplest, cheapest, and most compact reactor geometry of any magnetic fusion confinement configuration.

ZaP Energy is focused on deploying its flow Z-pinch reactor for fusion electricity and space propulsion [41]. The company has been awarded three ARPA-E funding grants to develop its concept, with the first going to the research team at the University of Washington in 2015, and the latter two going to the company after it span out. In 2020, the company announced that it had received funding it from the US oil and gas heavyweight, Chevron.

7.3.16 Marvel Fusion

Marvel fusion, based in Munich, Germany, was founded in 2019. Marvel Fusion's approach is based on a fast ignition ICF approach that intends to exploit the same progress in CPA laser technology as pursued by HB11 Energy. The Marvel Fusion approach, although details remain scant, envisages a two-stage laser ignition process akin to 'fast ignition' (refer to the description in chapter 2). The company will use a CPA laser coupled with a high compression laser to strike a target to achieve what the company says will be a 'neutron free' fusion reaction [42]. While it is not explicitly stated, this implies the company is pursuing an aneutronic fuel cycle. The company says it will build and operate a petawatt CPA laser system and will develop a sophisticated target in its development of a commercial fusion power plant. Marvel believes their reactor will be reduced in size and cost compared to existing ICF reactor concepts.

The company is supported by BlueYard Capital, a German venture capital firm and is currently assembling a team of scientists and engineers to develop its reactor from concept to reality.

7.3.17 Renaissance Fusion

Based in Grenoble, France, Renaissance Fusion was one of the first fusion start-ups in continental Europe, and the first private company focused on the stellarator[17]. The company was co-founded in 2019 by Francesco Volpe, an experimental stellarator scientist, and Martin Kupp, an academic with expertise in entrepreneurship. Renaissance is focused on developing the stellarator concept—which offers inherent plasma stability compared with other MCF reactors, but which is harder to build due to the complex geometries (see chapter 2)—while also exploiting advances in HTS magnet technology [43]. If successful, they will be the first to operate a high-field stellarator in the world, which the company hopes will accelerate their concept towards being the first commercial fusion power plant.

[17] There is now an additional private organization in the US, focused on stellarators, called Type One Energy, which came to our attention during the final stages of writing this chapter.

The company has not disclosed details of its current investment. However, several prominent fusion scientists and business associates as advisors who will support their mission are listed on the company's website.

7.4 Discussion: the prospect of private fusion success

While this chapter provides a useful summary of the majority of private fusion companies that exist today, it also shows the differences between the wide range of concepts. It is the authors' opinion that it seems probable that if fusion is to have a commercial future, then at least one of these companies will see its technology or its approach to innovation adopted by the new fusion industry. The prospects for such success are motivating the risk-tolerant investment that is needed to support this outcome. If, for example, one company emerges victorious in this way, one can expect that some of the other companies will fold, but this might be less common than one might expect at first impression. Indeed, the opposite could be true. Success in one fusion approach may cause a flood of investment into the sector to develop the best strategy, in the knowledge that fusion energy is itself viable as a commercial exploit. It could be that companies close to but not perfectly aligned with the emergent dominant design will redirect and adjust to find a role in a rapidly expanding technological ecosystem. Even companies dedicated to very different approaches may find that they have something special to offer. As the firms reorientate themselves, progress will accelerate.

All companies have a good understanding of the technical challenges right in front of them. These are, inter alia, those challenges that are detailed in chapter 2 of this book. They will—and must—develop a better understanding of market issues shaping the prospects for fusion as a commercial activity (see chapter 10).

As large established companies seek to move into the fusion space, it can be cost-effective and quick to acquire a fusion company whose own ideas were a near-miss. Such thinking allows us to suggest that all the fusion companies listed here (and several others featured elsewhere in the book, or those not mentioned at all) represent interesting endeavours that should receive greater attention and appropriate scrutiny[18].

In the future, we imagine many companies will fail, while one or two might achieve a breakthrough, and several might do well by other means. Such cognate opportunities include generating IPR for aspects of technology development, perhaps associated with commercial opportunities beyond fusion. We note that the Wright Brothers' 1903 Flyer does not in any way resemble a modern jet aircraft or even the planes of the First World War [44]. However, in successfully demonstrating that flight was possible, the Wright Brothers—whilst running their bicycle shop on the side—sparked a whole new technological era, which in the early years at least was closer to the technology of bicycles than modern jet aircraft. Picking a winner, which is not possible right now, is not the issue at hand. What matters is that, collectively, these private companies can be the pioneers that usher in the new technological era of commercially viable fusion

[18] Nothing in this chapter represents any kind of investment advice.

energy. Investment into the fusion sector now might allow the development of one or many of these concepts to move humankind towards commercialisation. If they are successful, investors, entrepreneurs and scientists alike have the potential to unlock a very big prize indeed.

References

[1] Pearson R J, Costley A E, Phaal R and Nuttall W J 2020 Technology roadmapping for mission-led agile hardware development: a case study of a commercial fusion energy start-up *Technol. Forecast. Soc. Change* **158** 120064

[2] The Economist 2019 Fusion power is attracting private-sector interest *The Economist*

[3] Singer P L and Bonvillian W B 2017 'Innovation Orchards': helping tech start-ups scale Washington, DC (http://www2.itif.org/2017-innovation-orchards.pdf)

[4] Gallagher K S, Holdren J P and Sagar A D 2006 Energy-technology innovation *Annu. Rev. Environ. Resour.* **31** 193–237

[5] Bonvillian W B and Weiss C 2015 *Technological Innovation in Legacy Sectors* (New York City: Oxford University Press)

[6] McCurdy H E 2001 *Faster, Better, Cheaper: Low-Cost Innovation in the U.S. Space Program* (Baltimore, MD: The Johns Hopkins University Press)

[7] Gaddy B E, Sivaram V, Jones T B and Wayman L 2017 Venture capital and cleantech: the wrong model for energy innovation *Energy Policy* **102** 385–95

[8] Volberg R 2016 The inevitable industry *American Security Project* (https://american-securityproject.org/the-inevitable-industry/)

[9] Rothrock R 2016 What's the big idea? *Issues Sci. Technol.* **32** 63

[10] Holland A 2020 Fusion development is critical to America's future. Congress should encourage public-private partnerships to accelerate development of fusion energy *American Security Project* http://americanfusionproject.org/fusion-recommendations-to-congress/

[11] Walker S P and Haines M G 1997 An assessment of the prospects for economic DT tokamak fusion, *Fusion Eng. Des.* **36** 387–405

[12] Brunner D 2018 Magnetic confinement fusion energy systems on the path to commercialization *IAEA Workshop on Fusion Enterprises* https://nucleus.iaea.org/sites/fusionportal/SharedDocuments/Enterprises/2018/Presentations/15.06/Brunner.pdf

[13] Guo H Y *et al* 2015 Achieving a long-lived high-beta plasma state by energetic beam injection *Nat. Commun.* **6** 1–6

[14] Binderbauer M W *et al* 2016 Recent breakthroughs on C-2U: Norman's legacy *AIP Conf. Proc.* **1721** 30003

[15] Baltz E A *et al* 2017 Achievement of sustained net plasma heating in a fusion experiment with the optometrist algorithm *Sci. Rep.* **7** 1–7

[16] Laberge M 2016 Acoustically driven magnetized target fusion at general fusion *Curr. Trends Int. Fusion Res. Proc. 7th Symp. Present.* **1154** 282–8

[17] Laberge M *et al* 2013 Acoustically driven magnetized target fusion *2013 IEEE 25th Symp. on Fusion Engineering (SOFE)* 1–7

[18] Lerner E J, Murali S K and Haboub A 2011 Theory and experimental program for pB 11 fusion with the dense plasma focus *J. Fusion Energy* **30** 367–76

[19] Fusion L P P 2020 LPP Fusion Website (https://lppfusion.com/)

[20] Austin-Morgan T 2019 An Oxford-based fusion project is closing on producing gain using a cheaper method than the mainstream *Eureka!*

[21] Ringrose T J *et al* 2017 A hypervelocity impact facility optimised for the dynamic study of high pressure shock compression *Procedia Eng.* **204** 344–51
[22] Proton Scientific 2020 Proton Scientific Website (http://protonscientific.com/history)
[23] Helion Energy 2020 Helion Energy Website (http://helionenergy.com/)
[24] McGuire T 2016 Overview of the Lockheed Martin compact fusion reactor (CFR) T4B experiment *APS* NP10–180
[25] McAdams R 2014 Beyond ITER: neutral beams for a demonstration fusion reactor (DEMO) *Rev. Sci. Instrum.* **85** 02B319
[26] McGuire T J 2018 Magnetic field plasma confinement for compact fusion power *Google Patents*
[27] Martin L 2020 Compact fusion – Lockheed Martin SkunkWorks website (https://lockheedmartin.com/en-us/products/compact-fusion.html)
[28] McGuire T J 2018 Encapsulating magnetic fields for plasma confinement *Google Patents*
[29] Sutherland D A and Jarboe T R 2019 Projected scaling of sustained spheromak configurations for economical fusion power *61st Annual Meeting of the APS Division of Plasma Physics* **64** UP10.00115
[30] CTFusion 2020 CTFusion Website (https://ctfusion.net/)
[31] Agni Fusion 2020 Agni Fusion Website (https://agnifusion.org/)
[32] Coppi B *et al* 2015 Perspectives for the high field approach in fusion research and advances within the Ignitor Program *Nucl. Fusion* **55** 53011
[33] Sorbom B N *et al* 2015 ARC: a compact, high-field, fusion nuclear science facility and demonstration power plant with demountable magnets *Fusion Eng. Des.* **100** 378–405
[34] Compact Fusion Systems 2020 Compact Fusion Systems Website (https://compactfusionsystems.com/)
[35] Hsu S C *et al* 2012 Spherically imploding plasma liners as a standoff driver for magneto-inertial fusion *IEEE Trans. Plasma Sci.* **40** 1287–98
[36] Lynn A G, Merritt E, Gilmore M, Hsu S C, Witherspoon F D and Cassibry J T 2010 Diagnostics for the plasma liner experiment *Rev. Sci. Instrum.* **81** 10E115
[37] Thio Y C F *et al* 2019 Plasma-jet-driven magneto-inertial fusion *Fusion Sci. Technol.* **75** 581–98
[38] HyperJet Fusion Corporation 2020 Hyperjet Fusion Website (http://hyperjetfusion.com/)
[39] Hora H *et al* 2017 Road map to clean energy using laser beam ignition of boron-hydrogen fusion *Laser Part. Beams* **35** 730–40
[40] Zhang Y *et al* 2019 Sustained neutron production from a sheared-flow stabilized Z pinch *Phys. Rev. Lett.* **122** 135001
[41] ZaP Energy 2020 ZaP Energy Website (https://zapenergyinc.com/)
[42] Marvel Fusion 2020 Marvel Fusion Website (https://marvelfusion.io/)
[43] Renaissance Fusion 2020 Renaissance Fusion Website (https://stellarator.energy/)
[44] Kline S J and Rosenberg N 1986 An overview of innovation *The Positive Sum Strategy: Harnessing Technology for Economic Growth* ed R Landau and N Rosenberg (Washington, DC: National Academy Press) pp 275–306
[45] Etzkowitz H and Leydesdorff L 1995 The Triple Helix—University-industry-government relations: a laboratory for knowledge based economic development *EASST Rev.* **14** 14–9

Section III

Public sector push to commercialization

IOP Publishing

Commercialising Fusion Energy
How small businesses are transforming big science
William J Nuttall, Satoshi Konishi, Shutaro Takeda and David Webbe-Wood

Chapter 8

STEP—on the pathway to fusion commercialization

Howard Wilson, Ian Chapman, Tris Denton, William Morris, Bhavin Patel, Garry Voss, Chris Waldon and the STEP Team

8.1 Introduction

Fusion combines light nuclei to create a heavier one and release energy that can be harnessed for electricity generation or other applications. The nuclear process is understood well, but creating the conditions on Earth that enables it is not. This is the technological challenge of reproducing fusion. Most approaches focus on the reaction between two isotopes of hydrogen—deuterium and tritium—to produce a helium nucleus and a neutron, each carrying large kinetic energy. It is this energy we seek to harness. Deuterium is plentiful, while tritium must be manufactured, probably *in situ* via breeder blankets.

While significant technical issues remain to be addressed, fusion energy is transitioning to the delivery era. This transition is being accelerated by the urgent need to provide energy to power industry and homes without incurring a detrimental impact on the global climate. Thus, private fusion companies are emerging around the world, proposing aggressive timelines that drive fusion much faster than typical government programmes have done in the past. They attract £Bns of private investment—a development that recognizes the market potential but, as we discuss in section 8.3, the gap to commercialization, and the associated challenges, are substantial. These private fusion companies are therefore more likely to succeed at this stage if they can benefit from the breadth of capability that government-funded fusion programmes provide. This chapter describes the UK strategy towards fusion commercialization via the STEP programme—Spherical Tokamak for Energy Production. Led by the UK Atomic Energy Authority, UKAEA, STEP aims to progressively grow industrial and academic partnerships to build supply chain capability, alongside the required fusion research and development (R&D), towards power plant concepts, figure 8.1.

doi:10.1088/978-0-7503-2719-0ch8

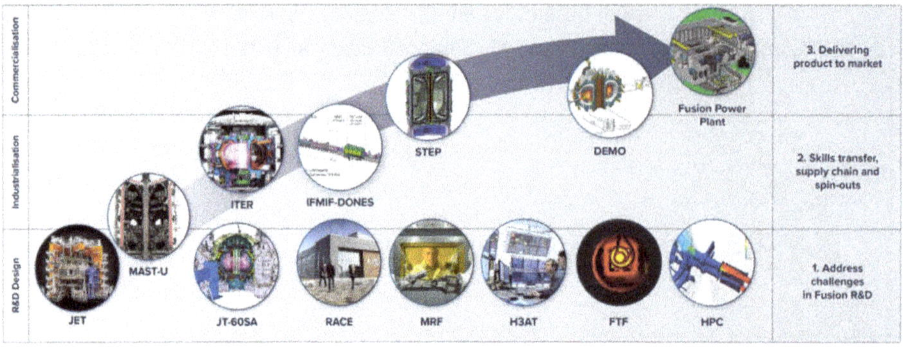

Figure 8.1. The UKAEA pathway to fusion commercialization, building on a suite of R&D facilities that span the breadth of fusion disciplines and taking advantage of the important roles of ITER and IFMIF-DONES – can STEP ease and accelerate the progress to a fusion power plant, complementing the DEMO approach?

UKAEA is the government body which has led the UK's national fusion programme since the huge potential for effectively limitless, safe, clean energy was realized, back in the 1950's. It has always been at the forefront of fusion energy research, embedded as a key player in a stimulating and internationally collaborative programme that has seen great advances over more than half a century. Specifically, it is a member of the EUROfusion consortium and involved in the ITER project [1, 2], as well as having strong links to the fusion programmes in the US, Asia and Russia. EUROfusion has developed a roadmap towards fusion commercialization [3], which at a high level aims to:

1. Address the science and technology questions on today's fusion facilities (including advanced computing);
2. Exploit new experimental facilities as they come on line, including the JT-60SA tokamak which is nearing completion in Japan; the international burning plasma experiment, ITER, and the materials irradiation facility, IFMIF-DONES [4];
3. Build and operate a demonstration power plant, DEMO, which is being designed now.

This approach takes ambitious steps between stages: DEMO engineering design will be essentially completed before high fusion power and breeder blanket testing in ITER and with limited data from IFMIF-DONES (further information from ITER and IFMIF-DONES will be used during construction and operation). This leads to a DEMO operating in the 2050's. Given the urgency and primacy of tackling climate change, this then raises the question of whether there are alternative pathways that can be pursued in parallel—specifically, can we speed up the commercial delivery of fusion power and/or reduce costs by introducing additional elements that complement and further enhance the strategy followed by EUROfusion? This is the question that UKAEA sets out to address with its STEP Programme.

8.2 Three components towards fusion commercialization

One can identify three components on the pathway to fusion commercialization, as follows (see figure 8.1).

Research and Development. In this first phase, we must address the science and technology questions associated with fusion power, making use of an extensive range of complementary experimental and computational capabilities across the world. Part of the complexity of fusion energy lies in the breadth of research disciplines it involves, bringing together the science of plasmas, materials and tritium, with engineering disciplines such as robotics, fluid dynamics, thermo-hydraulics, nuclear systems and power management. As well as extensive international collaborations, the UK hosts a suite of national facilities to study many of the critical aspects [5]: the MAST-U tokamak [6] provides a flexible fusion-relevant plasma physics base; the centre for Remote Applications in Challenging Environments (RACE) [7] develops and tests maintenance and inspection solutions; the Materials Research Facility (MRF) [8] enables testing and study of small activated material samples; the Hydrogen-3 (tritium) Advanced Technology (H3AT) centre [9] provides tritium science experiments, and the Fusion Technology Facilities (FTF) [10] provides a suite of experimental rigs to explore a range of manufacturing challenges. These capabilities, together with access to advanced computing, underpin a rigorous R&D programme to advance each of the fusion disciplines. However, the biggest challenge of fusion is widely recognized to be integration, so these individual disciplines need to be tackled in that context. For fusion plant designs, this integration is achieved by modelling, supported by practical experience on existing facilities. UKAEA operates the most advanced fusion reactor in the world—JET—for the European fusion science community. JET presently provides the best operational experience of integration anywhere in the world, completing a world-class fusion R&D capability in the UK.

Industrialization. This phase, which the field has now entered, progressively brings industry into the programme. This is important for two reasons: (1) it is industry that will build the first commercial reactors, so it must grow the skills and capacity to do this, and (2) to inform designs, ensuring manufacturability. It is ITER that has taken fusion into this industrialization era, with large scale fusion components manufactured by the world's companies and delivered to site. Another opportunity will be provided by the planned IFMIF-DONES £Bn-level fusion materials irradiation facility expected to operate in Europe by the end of the 2020s. The construction of smaller scale national facilities also drives an industrial interest, again helping to establish the supply chain required for the next step—commercialization.

Commercialization. In this phase, industry takes over as the main driver of the fusion programme, designing, building and commissioning fusion reactors for the market—an international market that is potentially worth £Tn's. EUROfusion's strategy is to enter the commercialization stage via DEMO, to a fleet of fusion power plants beyond that. There is a risk, well recognized in Europe, that the break between ITER and DEMO construction is too long to maintain the interest and momentum of industry, resulting in a loss of industrial capability between the two.

Furthermore, ITER is being built by the *world's* industry—each of the seven partners worldwide provides components as in-kind contributions. National programmes must grow within this international fusion landscape in order to drive industrialization towards commercialization without gaps in capability. This is recognized by China, for example, who, as well as partnering in ITER, are planning the Chinese Fusion Engineering Test Reactor (CFETR) [11]. Focussing on technology, it will complement ITER; however, another major benefit of CFETR is that its construction will maintain the momentum of fusion supply chain growth in China.

ITER, DEMO and CFETR are all large, conventional aspect-ratio tokamaks which will have a considerable capital cost, comparable to the large pressurized water fission reactors under construction in the UK and elsewhere [12]. This could impede the pace of penetration of fusion into the energy market, reducing its potential to help address climate change. Therefore it is worth asking whether we can reduce the cost of fusion power plants and hence accelerate the programme to fusion commercialization.

An efficient programme must recognize that the above three components are not independent. The technical R&D priorities, for example, must be informed and driven by the requirements of the industrialization and commercialization components in order to ensure a commercially viable product is developed. Thus, while the emphasis on each component shifts as the programme progresses, their needs must all be considered at all stages, and the STEP Programme is designed to enable that.

8.3 The role of private fusion companies

Private fusion companies are key to fusion commercialization. Their focus on an aggressive timeline, accepting and managing the higher risk, complements other approaches, driving a healthy innovation ecosystem and progress towards product delivery. However, they do not currently have the resources to invest in the full spectrum of R&D and capabilities mentioned in section 8.2 that are required to bring fusion to market. Nevertheless fusion is realizing private investment at a very significant level, typically driven by specific aspects of the full integrated fusion reactor design; these will often have near term societal and/or economic benefits in adjacent fields. This private approach to fusion commercialization is especially effective if linked to a wider landscape of R&D that manages the fully integrated reactor design, providing a framework and context for specific elements being advanced by fusion companies. While such a landscape has been created by international consortia such as EUROfusion or ITER, it is also important that national fusion programmes support the emergence and growth of private fusion companies, which in turn have a beneficial impact on the journey towards commercialization. This integrated approach, which establishes relations between private and public fusion, can increase the impact of the national investment, while also supporting competition, faster timescales and even greater innovation.

8.4 STEP—spherical tokamak for energy production

In recognition of the above requirements, UKAEA proposed the STEP Programme to the UK Government in 2018. This programme builds on a national strength in the compact fusion reactor design called a spherical tokamak, with UKAEA pioneering applications as both a fusion neutron source for science and technology [13, 14] and for electricity production [15, 16]. The promise that these early designs showed motivated and informed the recent £55M MAST Upgrade at Culham and, importantly, seeded the formation in 2009 of the company now known as Tokamak Energy. The complementary combination of national laboratory and a range of private industries driving fusion forward via the spherical tokamak will be a powerful one, bringing the benefits described in the previous section; unprejudiced partnership with industry is an explicit aim of the STEP Programme.

STEP is a national fusion programme that has mutually beneficial synergies with EUROfusion activities, aiming to drive the UK's ambition to be amongst the first to deliver fusion power. The headline aim of the STEP Programme is to design, build and operate the STEP Prototype Reactor (SPR) by 2040 that will deliver net electricity at the level of 100 MW(e), demonstrating and informing many aspects of a commercial reactor fleet to follow (see figure 8.2). However, the full Programme does much more than that, and is designed to:

- Deliver outputs to help inform a fusion regulatory framework
- Stimulate commercial investment
- Innovate, creating solutions that find near term applications in adjacent sectors
- Stimulate growth of the fusion energy supply chain through partnering
- Nurture skills in a diverse and inclusive way, training those who will deliver fusion power and supporting skills growth in adjacent sectors
- Support industry to develop designs for a first commercial fleet of fusion reactors to follow the SPR
- Develop the new STEP site and associated infrastructure

STEP is an adventurous programme, and there is no guarantee of success. The biggest risks are scientific and technical ones—identifying the solutions to contain and sustain a burning fusion core. To manage this high risk, the programme is divided into three tranches:

- Tranche 1 (2019–2024) will deliver an integrated concept design for the prototype reactor, and a clear strategy to address any remaining technical risks; it will also identify a UK site and establish the operational framework for subsequent tranches.
- Tranche 2 (2024–2032) will develop the engineering design of SPR, testing and optimizing different options for the subsystems; this period will see significant development of the STEP site.
- Tranche 3 (2032–2040) will see the construction and commissioning of SPR, ready to deliver its net fusion-powered electricity.

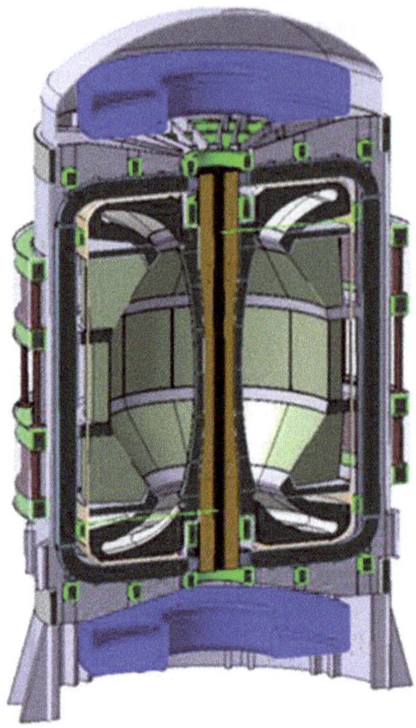

Figure 8.2. Early design option for the STEP Prototype Reactor, SPR, which aims to deliver in the region of 100 MW of electricity from fusion power.

Tranche 1 has received £222M of public funding to realize a concept design. While public funding is appropriate in Tranche 1, the programme will become increasingly commercial as it progresses, seeking significant commercial investment for subsequent tranches. This places a different emphasis on the approach from traditional fusion R&D—STEP must have in place an investible concept and programme by the end of 2023. To achieve this, the programme adopts a more commercially-minded approach, working with industry, national laboratories and academic partners to generate and manage its intellectual property (IP) throughout Tranche 1. Without a portfolio of IP, STEP is unlikely to be investible, with consequences for commercial engagement beyond Tranche 1.

It is worthwhile reflecting on how STEP supports and complements the emergence of fusion in the private sector. It does this in several ways:

- It provides a suite of enhanced research facilities to address technical problems, both for STEP and for external companies;
- It provides an integrated design framework that can either inform the approach of private fusion activities, or receive specific components or subsystems designed by them;
- It will inform and drive the development of a fusion regulatory framework;
- It will expand the capabilities of industry and the associated supply chain;
- It will grow the national fusion skills base.

Combining private and public approaches—in a balance that will evolve over time—will deliver a strong national capability that will underpin a drive towards realizing fusion power on the grid.

While STEP is necessarily a national programme, it will not succeed without strong international collaborations. STEP will have more similarities with other reactor designs than differences, and it is important to collaborate on these to the benefit of both partners. This will include most features of plasma physics, as well as technologies such as maintenance, tritium breeding, the fuel cycle, power management, exhaust heat load and materials properties under high neutron fluence. ITER, for example, is a key element of the STEP strategy for four main reasons: (1) seeding the supply chain is already under way; (2) a demonstration of manufacture, installation and integration of the fusion plant systems; (3) learning how to evolve from first plasma to full fusion performance with the deuterium–tritium fuel mix, and (4) burning plasma scenario development and physics. Like the European DEMO, STEP will need more information than ITER alone will provide. For example, ITER will not provide sufficient neutron fluence to adequately test materials. Europe and other nations already conduct extensive and valuable materials tests following irradiation in fission test reactors, noting that while fusion neutrons are much more energetic than fission neutrons when first produced, they slow down through interactions with the reactor structure so that a fission spectrum is relevant for many areas. Nevertheless, there are additional mechanisms at play for material interactions with higher energy neutrons—relevant for components facing the plasma—and so Europe plans to construct and operate an accelerator-based neutron source called IFMIF-DONES by the end of the 2020s. This will expose a large number of small material samples to sufficient neutrons to enable rigorous tests of their performance following reactor-relevant irradiation. The IFMIF-DONES facility, including collaborations in the associated experiments, will help STEP by confirming materials choices for SPR and the subsequent commercial fleet, as well as informing their performance/maintenance requirements.

ITER and other facilities planned worldwide are important to STEP, and have completion timelines beyond 2024; this highlights the fact that we will not have all the information we need by the end of Tranche 1. Specifically, significant technical risks will remain as we transition into the engineering design activity of Tranche 2, and even into the construction phase of Tranche 3. This is the risk that STEP accepts, and seeks to manage, in driving a faster programme—to wait for all information results in delays to the delivery. Digital engineering will play a significant role in mitigating this risk. The foundations for this will be laid during Tranche 1, understanding the capabilities and scale required for Tranche 2 and beyond. The main features of a strong digital engineering programme for STEP will include:

- In-silico models of the plasma and selected components, into which we can insert additional knowledge as it appears (such as materials behaviours and failure modes).

- Uncertainty quantification—what is the uncertainty in the performance of a sub-system that arises due to lack of data and gaps in knowledge?
- Uncertainty propagation—how does uncertainty in the different subsystems influence the overall device performance, availability and maintenance strategy?
- Design optimization—a fusion reactor is a complex system, with most subsystems interacting with most others; how does one understand these interactions and develop an optimized overall design?
- Digital twinning—developing a digital twin of the reactor system and/or subsystems to understand how performance varies with time as materials, for example, degrade under neutron irradiation.

This final point is a particularly interesting feature of designing a fusion reactor. Because the material's properties evolve under high energy neutron irradiation, the behaviour of the device will change with operational time. A key ambition is to quantify this and so inform a maintenance, availability and lifetime strategy which is likely to be an important input to a fusion regulatory framework and, ultimately, the cost of electricity—key issues for a stakeholder justification.

8.5 Fusion power from a spherical tokamak plasma

Following the pioneering experiments on START [17], many spherical tokamaks (STs) were built around the world, including the two largest—MAST [18] in the UK and NSTX in the US [19], both of which are currently undergoing significant upgrades to capability. STEP will employ this promising spherical tokamak design, which has been explored previously for fusion applications [13–16, 20]. A tokamak fusion reactor confines the fuel mix, typically deuterium and tritium, in plasma form by a toroidal geometry of magnetic field. A spherical tokamak is a compact version, minimizing the radius of the hole at the centre of this geometry. If the plasma of a conventional, JET-like, tokamak can be likened to a ring doughnut, that of a spherical tokamak is like a cored apple. Topologically, a spherical tokamak is a torus, but it has the overall appearance of a sphere.

Figure 8.3 shows the essential features of the spherical tokamak plasma core (the principles are common to all tokamaks). At its heart is a central rod which carries a large current—the rod current. This creates the so-called toroidal component of magnetic field, but this alone is insufficient to confine the plasma. An additional poloidal component of magnetic field is created by passing a current through the plasma. The resulting magnetic field lines spiral around a set of (imaginary) nested toroidal 'flux' surfaces within the plasma. Additional coils (not shown) surround the plasma to shape and control it.

While it poses some plasma and engineering challenges, an immediate benefit of the spherical tokamak is its compact nature—it has a smaller radial build than a conventional tokamak, which is expected to reduce the capital cost, as well as the initial tritium charge and operational inventory. The design around a central rod brings potential manufacturing advantages related to modularity, especially if

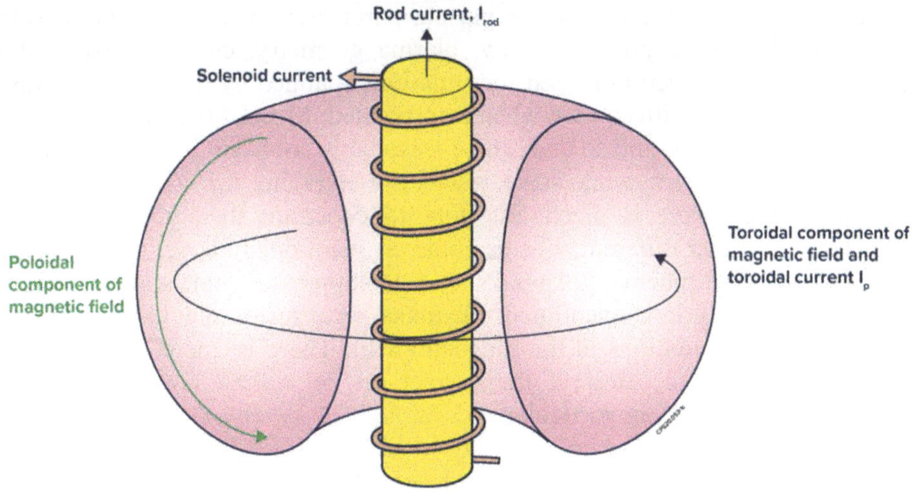

Figure 8.3. Compact magnetic field geometry of a spherical tokamak showing the rod current, I_{rod}, to provide the toroidal component of magnetic field; the toroidal plasma current, I_{p}, to provide the poloidal component of the magnetic field, and the solenoid used in present spherical tokamaks to induce the toroidal current in the plasma.

de-/re-mountable joints can be developed for the magnets; this then opens up new options for remote maintenance.

A critical plasma physics feature is associated with the amount of plasma pressure that can be contained with a given magnetic field. This is measured by an efficiency parameter called beta (β), which is the ratio of the thermal energy density stored in the plasma to the energy density stored in the confining magnetic field. A spherical tokamak can accommodate much higher values of beta than a conventional tokamak because of the high ratio of plasma current to magnetic field it can sustain [21]. Simulations indicate that high beta is beneficial for suppressing some micro-instabilities that drive turbulence, thus providing a potential route to improved confinement relative to conventional tokamaks—this is likely critical for the ST to be viable. However, other micro-instabilities are predicted to become more prevalent at high beta, so whether or not there is a net benefit—perhaps at a 'sweet spot' in beta—remains to be demonstrated. Confinement is quantified by the energy confinement time—the time taken for $1/e$ ($e = 2.718$) of the thermal energy to leak from the plasma if the heating power is removed. The more turbulent the plasma, the faster the energy leaks out and the shorter the confinement time—understanding the turbulence and the instabilities that drive it is therefore important for identifying plasma scenarios with sufficient confinement.

In the present absence of a predictive, first-principles model of turbulence for spherical tokamak plasmas, one approach to estimate the confinement time is to use scaling laws fit to a large dataset derived from very many experiments on the world's tokamaks. A scaling law represents the confinement time as a product of parameters, each raised to a particular index. These indices are determined by fitting to the tokamak database. The result is an equation—the scaling law—that describes how the confinement time depends on the plasma parameters, such as plasma current, magnetic field, heating power, density, plasma geometry, etc. It is expected to provide a reliable prediction if used for situations spanned by the original dataset, but should be treated with caution when extrapolated. Most of the data comes from conventional tokamaks, and so one can be reasonably confident in using the scaling law for ITER. However, while some results are emerging for spherical tokamak scaling laws [22], there is much less data to constrain them and the plasma parameters are rather different. For example, as mentioned above, the high value of beta that exists in spherical tokamaks tends to suppress the fine scale instabilities that drive turbulence in conventional tokamaks, but creates new turbulence drive mechanisms via fluctuations in the magnetic field. There is, therefore, significant uncertainty in how to interpret the predictions of scaling laws for spherical tokamaks, and the SPR, in particular.

To provide a specific example, we introduce in table 8.1 a particular plasma scenario for a 810 MW fusion power plasma core. The major and minor radii, R and a, are selected to provide sufficient space to shield the centre column from the neutrons and other radiation—these shielding requirements vary with magnet conductor choice, but are all rather uncertain in the absence of high energy neutron irradiation data. The elongation is the ratio of the height of the plasma to the diameter of its poloidal cross-section, measured across the mid-plane (i.e. $2a$). A high value of elongation is beneficial as it provides access to high fractions of the so-called 'bootstrap current'. This bootstrap current flows in all hot tokamak plasmas, and is proportional to the pressure gradient; a high value of this reduces the power requirements from the auxiliary current drive system, improving the reactor efficiency. Thus, for our example equilibrium, only 3.3 MA out of the total of 16.5 MA of plasma current needs to be driven by external means; theory predicts that a 60 MW neutral beam injection system can provide this, for example. The rod current provides the main, toroidal confining magnetic field—we have pushed it to

Table 8.1. Example parameters for a spherical tokamak fusion plasma core.

Parameter	Value
R/a (m)	2.5/1.5
Elongation	2.8
Triangularity	0.55
Plasma/Rod current (MA)	16.5/30
Driven current (MA)	3.3
β(%), β_N	25, 5.4
n/n_{GW}	0.9
$T_{i0} = T_{e0}$ (keV)	17
Safety factor on axis	>2
Fusion power (MW)	810
External heating power (MW)	60

be rather high here in order to achieve an acceptable value for the normalized pressure, β_N. This is the ratio of beta (β, expressed as a percentage) to the normalized current—plasma current (in MA) normalized to the product of magnetic field (T) and minor radius (m). The value of β_N is limited by large scale ('MHD') plasma instabilities. The constraints from these can be reduced, allowing higher β_N, by raising the central safety factor, q_0, above 2 [23], noting that q_0 is inversely proportional to the ratio of current density at the plasma centre to the magnetic field there. Thus, we seek to drive current away from the magnetic axis, which also helps us to achieve the high elongation required (indeed, this requirement aligns well with a high bootstrap current fraction, which naturally drives current off-axis). The average β of this example is 25%, which is an order of magnitude higher than that planned for the conventional tokamak, ITER, of about 2.5%. The plasma density, n, is normalized to the Greenwald limit (which in turn is related to the average current density of the plasma); in a conventional tokamak, this ratio is typically maintained below unity to avoid high disruptivity plasmas.

Having explained the parameters of table 8.1, let us now return to confinement, and consider the results of three scaling laws for this example: the IPB98(y,2) ITER scaling law (equation (30) of [24]); one due to McDonald (see equation (40) of [24]), and one due to Petty (see equation (38) of [24]). We define a factor, H, which is the ratio of the energy confinement time required (1.1 s here) to the prediction of each scaling law. The scenario provided in table 8.1 requires $H_{\mathrm{IPB98(y,2)}} = 1.26$, $H_{\mathrm{McDonald}} = 1.02$ and $H_{\mathrm{Petty}} = 1.08$. Thus the confinement assumptions for this scenario are optimistic compared to the IPB98(y,2) scaling law, and broadly consistent with the other two. This variation in the predictions of the scaling laws illustrates the uncertainty in those predictions (a 20% enhancement of confinement is significant), and reinforces the need for a first-principles model for confinement, including the impact of magnetic field fluctuations on the turbulence.

Despite its uncertainty, confinement time is one of the biggest design drivers for a fusion power plant. Figure 8.4 illustrates the impact of improved confinement relative to the IPB98(y,2) scaling law on the current drive requirements. Thus, we

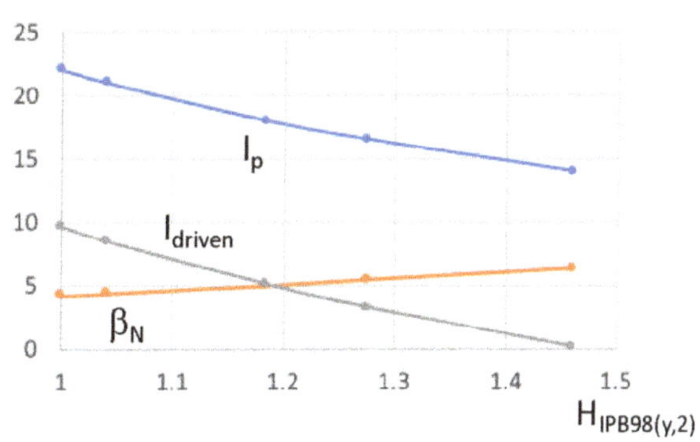

Figure 8.4. Variation of driven current, I_{driven}, required with confinement H-factor, using the scaling from IPB98(y,2). Also shown are the plasma current, I_p and normalized plasma pressure, β_N. Currents are in MA and the fusion power is maintained at 810 MW.

hold the fusion power and external heating power fixed at the values of table 8.1, and explore the consequences of being able to achieve different values of confinement enhancement factor, H, relative to the IPB98(y,2) scaling law. For higher values of H, figure 8.4 shows that we can reduce the plasma current, I_p, required to achieve the necessary confinement time, which then means the current to be driven by external systems can be lowered. This is important because it requires a lot of power to drive current in the plasma (non-inductively), with consequences for commercial viability. Indeed, this externally driven current, I_{driven} in figure 8.4, goes to zero for H just below 1.5. Conversely, at lower values of H, significant current must be driven by the external system which would then require much more external power than the 60 MW assumed here, with implications for whether or not net power is produced. Confinement is therefore key for commercial viability of tokamaks and, in particular, it remains a significant uncertainty for spherical tokamaks. MAST-U and NSTX-U in the US are expected to shed light on this key issue for STEP [18, 19], combined with advances in theory and simulation. In the private sector, ST40 [25], constructed and operated by Tokamak Energy, is designed to provide important data at high magnetic field.

8.6 Technical risks to commercial viability of a spherical tokamak

In this section, we identify the technical challenges that are likely to have the biggest impact on the timeline to fusion commercialization. While the focus of this chapter is on the spherical tokamak, most of the issues are generic to tokamaks and other magnetic confinement systems.

We have explored some of the physics challenges for the burning plasma core in the previous section, and these are amongst the most significant uncertainties for the spherical tokamak; however, there are others. The layout of a spherical tokamak core provided in figure 8.3 shows the central solenoid that is usually used to generate the current in today's tokamak plasmas, and is planned for the baseline operating regime for ITER as well as conventional tokamak fusion power plants with cyclic operation. In a fusion reactor, this solenoid would need to be shielded from the neutrons to avoid breakdown of the insulator. While this is relatively straightforward in a conventional tokamak, it is challenging to arrange in a compact ST due to the space constraints. Engineering solutions for this might be possible to enable a solenoid to assist transient ramping of the plasma current up to its steady state, and controlled ramp down to terminate the discharge; however a steady state ST plasma will likely require the identification of additional non-inductive current drive schemes to supplement the bootstrap current. Two possibilities are to use neutral beam injection [20] or electrostatic electron Bernstein waves (EBW) [26]. The former, used to provide the estimate of 60 MW in table 8.1, increases the nuclear island and requires large ports (with detrimental impact on tritium breeding), while the latter is unproven, with much uncertain physics. Electromagnetic microwaves at a frequency that resonates with the electron cyclotron frequency in the plasma (like EBW) provides another option, but only at high magnetic field, above that given in table 8.1. Such non-inductive current drive is also likely to be required to supplement the solenoid during ramp-up/down, and perhaps even replace it altogether.

With any compact fusion device, managing the heat exhaust over a relatively small area is a challenge. Innovative magnetic geometries have been proposed for this, including the snowflake [27] and the Super-X [28] divertors. In the Super-X, for example, a plume of exhaust plasma is drawn out to where substantial power can be radiated over a large area in an enclosed chamber, well away from the core plasma. The £55M upgrade to MAST-U is largely designed to optimize the capability of the Super-X (see figure 8.5), and to quantify its impact on core confinement [29].

The ability to breed sufficient tritium—more than the reactor burns to provide power—has never been demonstrated on any fusion reactor. Tritium breeding blanket modules must be positioned around the plasma to capture and multiply the neutrons and, via a reaction with lithium, produce the tritium. This tritium then has to be extracted, separated out from other gases and then processed before being injected back into the plasma. The kinetic energy of the neutrons is converted to heat in the blanket which can either be extracted to drive turbines for electricity generation, or used directly for other industrial processes (e.g. hydrogen manufacture, desalination, etc). While there are many generic issues associated with the blanket modules that cut across all fusion reactor designs, the spherical tokamak has a number of special features. The most obvious is the limited ability to accommodate blanket modules on the inboard side due to space constraints. At very tight aspect ratio, acceptable solutions exist with no inboard breeding as only a small fraction of neutrons are lost to the centre column. However, as the aspect ratio is increased this loss also increases, and at some point (above an aspect ratio of approximately 2) it becomes necessary to capture these neutrons, and also breed on the inboard side to achieve the required tritium production. Exploiting the space created by a

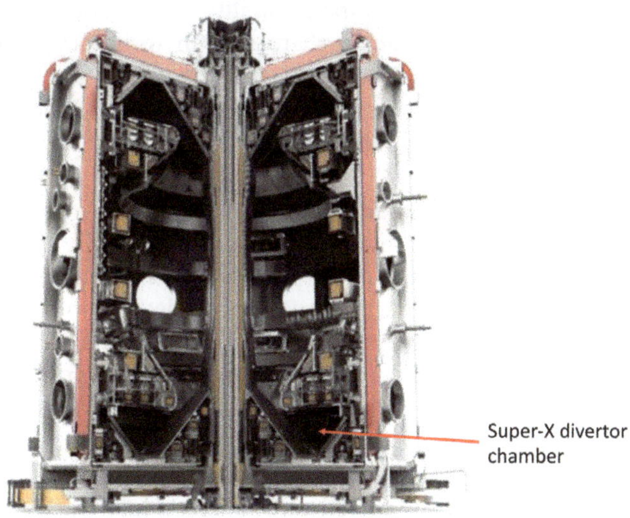

Figure 8.5. Cross-section through the new MAST-U spherical tokamak, highlighting the lower Super-X divertor chamber.

long-legged divertor can also help achieve the required tritium production [20]. The different solutions for tritium breeding have implications for maintenance, and hence availability which is a significant consideration for commercial viability.

The materials of the components that face the plasma, which include the breeding blanket modules, must withstand the hostile neutron environment. As mentioned above, this energetic neutron flux can change the physical properties of materials over time [30], which includes creation of radioactive isotopes via the nuclear reactions with the neutrons. Designing materials with the appropriate structural and/or functional properties as they evolve through operations, while at the same time optimizing the consequences for waste, is one of the greatest challenges facing commercialization of fusion. These material properties are also influenced by their permeability to tritium, influencing waste strategies as well as the tritium inventory. There are design and commercial implications arising from (robotic) maintenance, availability, waste management and licensing, which must be considered in the design from the very start. The challenge is exacerbated by the present lack of an appropriate neutron source for testing materials and, in the end, this requirement could determine the timeline to a commercial power plant. The interested reader is invited to consult the 2017 Special Edition of the IAEA Nuclear Fusion journal, introduced in reference [31], as well as a more recent review of the EU DEMO materials research strategy [32].

The ability to achieve high magnetic field with acceptable power loss is important for plasma stability, non-inductive current (e.g. using microwaves) and confinement. At the same time, the magnet system should be demountable if possible, to ensure efficient maintenance schedules and minimal downtimes—a key commercial driver. There are three options for conductor choices:

- Cryogenic, normal conductors have the highest power losses, but do not necessarily dominate the recirculating power and have proven solutions for demountable joints; an issue is the feasibility of the associated cryo-plant.
- Low temperature superconductors are proven technology, and have clear applications for the poloidal field coils; key questions include whether there exists a solution for the centre column where (1) their critical magnetic field and current density may constrain the toroidal field; (2) the shielding requirements are high pushing up the radial build, and (3) there is no proven solution for a demountable joint.
- High temperature superconductors show great promise, providing potential access to very high magnetic fields (provided an engineering solution can be found for the magnet system), and with possible solutions for demountable joints. Key issues include the timescale to commercial availability—will they be ready in time for the first demonstration fusion plants—and their susceptibility to neutron irradiation is less well known, leading to uncertainties in the amount of shielding required.

The commercial potential of high temperature superconductors (HTS) in fusion has underpinned significant private investment, both for spherical tokamak [33] and conventional tokamak [34] approaches, both pushing towards high magnetic field solutions. A possible strategy which STEP is exploring is to develop the first prototype reactor based on known technology, but ensure that designs can be upgraded to HTS when it becomes proven.

The breadth of the design challenge is large, spanning a range of science, technology and engineering. Furthermore, as we have already mentioned, a tokamak is a complex machine with many inter-connects that couple the different components. This requires an integrated approach, and therefore to maintain an aggressive timeline, the above technology and engineering challenges will need to be addressed alongside the development of the reactor plasma core design.

8.7 Conclusions

Fusion power is entering the industrialization era, and to maintain the momentum towards commercialization will require government laboratories to work with industry, including private fusion companies, towards demonstrating a viable, sustainable product for the energy market. The economic value will be huge, adding greatly to the tremendous societal and environmental benefits that fusion power will bring. However, there is a number of challenges to overcome on the path to commercialization of fusion energy, and this must be done sufficiently fast to deliver an impact on climate change, and benefit further from the emerging and growing interest from investors. The UK's STEP programme is a national programme that will work in synergy with strong international collaborations, while also seeking ways to partner with industry and academia to accelerate the commercial realization of fusion power. The focus will be on the spherical tokamak—a compact magnetic confinement fusion device that could deliver fusion power for lower capital cost. There remain significant science and technology challenges, but with innovation and creativity in design, STEP aims to show that an early demonstration of fusion-powered electricity is feasible.

Acknowledgments

This work was performed in part with funding from EPSRC grant number EP/R034737/1.

References

[1] Bigot B 2019 *Nucl. Fusion* **59** 112001

[2] See www.iter.org

[3] See www.euro-fusion.org/eurofusion/roadmap/

[4] Knaster J *et al* 2017 *Nucl. Fusion* **57** 102016

[5] Chapman I T and Morris A W 2019 *Phil. Trans. R. Soc. A.* **377** 20170436

[6] https://ccfe.ukaea.uk/research/mast-upgrade/

[7] www.race.ukaea.uk

[8] https://mrf.ukaea.uk

[9] https://ccfe.ukaea.uk/research/h3at/

[10] https://ccfe.ukaea.uk/research/fusion-technology-facilities/

[11] Zhuang G *et al* 2019 *Nucl. Fusion* **59** 112010

[12] www.edfenergy.com/energy/nuclear-new-build-projects/hinkley-point-c
[13] Hender T C, Voss G M and Taylor N P 1999 *Fusion Eng. and Design* **45** 265
[14] Voss G M *et al* 2008 *Fusion Eng. and Design* **83** 1648
[15] Voss G M *et al* 2000 *Fusion Eng. and Design* **51-52** 309
[16] Wilson H R *et al* 2004 *Nucl. Fusion* **44** 917
[17] Sykes A *et al* 1999 *Nucl. Fusion* **39** 1271
[18] Harrison J R *et al* 2019 *Nucl. Fusion* **59** 112011
[19] Menard J E *et al* 2017 *Nucl. Fusion* **57** 102006
[20] Menard J E *et al* 2016 *Nucl. Fusion* **56** 106023
[21] Gryaznevich M *et al* 1998 *Phys. Rev. Letts.* **80** 3972
[22] Buxton P F *et al* 2019 *Plas. Phys. Control. Fusion* **61** 035006
[23] Hole M J *et al* 2010 *Plas. Phys. Control. Fusion* **52** 125005
[24] Doyle E J *et al* 2007 *Nucl. Fusion* **47** S18
[25] www.tokamakenergy.co.uk/st40/
[26] Shevchenko V *et al* 2007 *Fusion Sci. Technol.* **52** 202
[27] Ryutov D D and Soukhanovskii V A 2015 *Phys. Plasmas* **22** 110901
[28] Kotschenreuther M *et al* 2010 *Nucl. Fusion* **50** 035003
[29] Morris A W *et al* 2018 *IEEE Trans. Plasma Sci.* **46** 1217
[30] Gilbert M R and Sublet J-C 2011 *Nucl. Fusion* **51** 043005
[31] Stork D and Zinkle S J 2017 *Nucl. Fusion* **57** 092001
[32] Pintsuk G *et al* 2019 *Fus Eng. and Design* **146** 1300
[33] Costley A E 2019 *Phil. Trans. R. Soc. A.* 37720170439
[34] Whyte D 2019 *Phil. Trans. R. Soc. A.* 37720180354

IOP Publishing

Commercialising Fusion Energy
How small businesses are transforming big science
William J Nuttall, Satoshi Konishi, Shutaro Takeda and David Webbe-Wood

Chapter 9

DEMO design activities and helical initiatives in Japan

Takuya Goto

In this chapter, the current situation of fusion research and development (R&D) in the public sector of Japan is reviewed. Currently in Japan, fusion energy is regarded as a new energy source that will be an alternative to fossil fuel rather than as next generation nuclear energy. The decision making regarding fusion energy development is being conducted by the Science and Technology Committee on Fusion Energy, which is placed under the Ministry of Education, Culture, Sports, Science and Technology of Japan (hereafter 'MEXT'). The background of this situation is reviewed and two major activities for the fusion DEMO reactor design based on magnetically confined systems, the tokamak and the helical, are introduced. Another important activity for fusion R&D, laser fusion, is described in chapters 2 and 6.

9.1 Japan's policy on fusion research and development

In Japan, the phased programme for fusion R&D has been formulated by the Atomic Energy Commission (AEC) of Japan. 'The Phased Basic Programmes for the Fusion Research and Development' have been defined by the Advisory Committee established under the AEC. Each basic programme has a specific goal and a core device. The first phase basic programme started in 1968. The goal was the development of an axisymmetric torus device with the plasma beta value (the ratio of plasma pressure and magnetic pressure, one of the important indexes of the performance of a magnetically confined plasma device) of ~1%. In this phase, JFT-2 tokamak in JAERI (Japan Atomic Energy Research Institute) was assigned as a core device. According to the achievement in the first phase, the second phase basic programme started in 1975. The goal of this phase was the achievement of the reactor-relevant plasma condition: core plasma temperature of 10 keV (~100 million °C) and the product of plasma density and confinement time $n\,\tau \sim 2\text{–}6 \times 10^{19}\ \mathrm{m^{-3}\,s}$. These parameters are important factors to prospect for the achievement of the

doi:10.1088/978-0-7503-2719-0ch9

Lawson condition, which corresponds to the fusion gain (the ratio of the energy produced by the fusion reaction to that injected into the device) of unity. In this phase, JT-60U tokamak in JAERI was assigned as a core device. Currently, the third phase basic programme, which started in 1992, is in progress. The goal is the achievement of a high fusion gain (~20) and a long pulse operation (~1000 s). At the time of the definition of the third basic programme, the core device of this phase was considered as a certain tokamak experimental reactor and two other devices were expected to be built before commercialization: a prototype reactor for the technical demonstration of electricity generation and a demonstration reactor for the economic demonstration.

In 2005, the third phase basic programme was reviewed and a report entitled 'Future Fusion Research and Development Strategy' was issued. In this report, the ITER was assigned as a core device of the third phase basic programme and it was defined so that the economic demonstration of a fusion power plant would be conducted by the next step reactor (DEMO reactor) and the related elemental technological development. The goal of the fourth phase has been defined as the construction and operation of the fusion DEMO reactor. This means that this fourth phase will be the last phase towards commercialization and the fusion DEMO reactor will be the last step towards the commercial fusion power plant. After this report was issued, the necessity to decide upon a strategic roadmap with clarification of the required technology has been noted in several reports issued by the Working Group on Fusion Research (which is established under the MEXT Council for Science and Technology) and the AEC's Advisory Committee on Nuclear Fusion. After many discussions, the Joint-Core Team for the Establishment of Technology Bases Required for the Development of a Demonstration Fusion Reactor (hereafter 'Joint-Core Team') was established in July 2013. The Joint-Core Team conducted the reconsideration of the basic concept required for the DEMO and the creation of the timeline chart that visualizes the evolution of the establishment of the technological basis required for the technological issues for the DEMO construction based on the tokamak system. These activities are summarized in two issued reports: 'Basic Concept of DEMO and Structure of Technological Issues' (July 2014) [1] and 'Chart of establishment of technology bases for DEMO' (January 2015) [2]. According to the discussion in the Joint-Core Team, the definition of Japanese fusion DEMO reactor (JA DEMO) has also been changed. The latest definition is as follows:

- Realization of a steady and stable electric output of over several hundred megawatts
- Achievement of the availability that is sufficient for commercialization
- Overall tritium breeding that fulfils self-sufficiency in fuel

The Joint-Core Team also provided a recommendation on the strengthening of the DEMO design activity and the establishment of the all-Japan organization that will effectively promote the activity and be shared by industry, government and academia.

After the Great East Japan Earthquake in 2011 and the nuclear power plant accident, the role of AEC was reviewed and MEXT was designated as the main organization that promotes Fusion R&D as part of science and technology policy. In March 2015, The Science and Technology Committee on Fusion Energy (hereafter 'Fusion S&T Committee') was established in MEXT by upgrading the Working Group on Fusion Research, which belonged to the Nuclear Science and Technology Committee. The Fusion S&T Committee has formulated 'Action Plan of Japan toward Development of DEMO reactor' [3] in February 2016 to classify the technological issues and to clarify the linkage between the issues. In December 2017, the Fusion S&T Committee formulated a new fusion R&D policy, 'Japan's Policy to promote R&D for a fusion DEMO reactor' [4], based on the latest progress in R&D and the change in the domestic and international situations including the delay of the ITER project. In these two reports, the approach for transitioning to the DEMO phase (the fourth phase basic programme) is clearly noted. The decision to transition to the DEMO phase will be taken in the 2030s when fusion operation (DT) of ITER is expected. It is necessary that the economic feasibility of a commercial reactor is foreseeable when transitioning to the DEMO phase. This does not necessarily mean that the DEMO reactor itself needs to have a high economic efficiency. The design of the commercial power plant can be changed from that of the DEMO reactor to improve economic efficiency. However, the capability for safe and economical operation of the commercial power plant needs to be confirmed through the operation of the DEMO reactor. Before the final decision to transition, two intermediate checks and reviews (C&R) will be scheduled. The first intermediate C&R is scheduled for around 2020, which is the timing of the beginning of the operation of JT-60SA tokamak. If the progress in R&Ds meets the designated target, the DEMO R&D shifts from the pre-conceptual design phase to the phase for conceptual design and elemental technology development. The second intermediate C&R is scheduled to coincide with the timing of ITER's first plasma, which is currently scheduled for 2025. After this C&R, DEMO R&D will enter the phase of engineering design and full scale technological development. The schedule is strongly affected by the ITER progress and the commercialization of fusion energy will be expected in the 2050s at the earliest.

In accordance to the roadmap, conceptual design activity of JA DEMO is ongoing. Before the formulation of the Action Plan, joint design activities with the EU have been conducted to address critical issues for the DEMO development under the framework of DEMO Design Activity (DDA) in Broader Approach (BA) activity, which is the complementary R&D project to be conducted in parallel with the ITER project for the construction of the fusion DEMO reactor. The Phase I of BA DDA terminated in March 2020, but this activity will continue as a part of the BA DDA Phase II. However, this activity is now regarded as a sharing and broadening of the knowledge of both partners because the approach to the DEMO is different in the EU and Japan. Both the EU and Japan consider the DEMO reactor to be the unique and last device before commercialization. However, the assumed operation scenario and type of breeding blanket are different. Specific design considerations are also required based on national laws and regulations.

Therefore, the conceptual design of JA DEMO is currently promoted by the domestic organization, the so-called 'Joint Special Design Team for Fusion DEMO' (hereafter as 'Joint Special Team'), which was established in June 2015 responding to the recommendation by the Joint-Core Team. Currently over 100 members have joined the Joint Special Team from institutes, universities, and companies all over Japan.

9.2 Status of tokamak DEMO design in Japan

In the past, various conceptual designs of tokamak fusion power plants have been proposed [5–10]. Most of them assume the adoption of several *advanced* concepts in both the core plasma physics design and the reactor engineering design to achieve a sufficient economic efficiency that is competitive with the existing large scale power plants. On the other hand, the conceptual design activity by the Joint Special Team has taken a different approach. The design activity started with the establishment of the single 'reference design' that is feasible based on the present physics and the technological basis. That means the design fully utilizes the technological basis of ITER and light water reactors. After the establishment of the reference design, the design activity will move on to the improvement of the reference design in order to solve the design issues specific to the fusion DEMO. For this reason, the primary design parameter of JA DEMO is relatively conservative compared with the past power plant designs.

Considering the plant economic efficiency, the power density of the reactor (the ratio between the thermal or electric output to the reactor size) should be maximized. Consequently, most of the past designed reactors have a similar thermal output as a large scale power plant (~3 GWe) with a smaller reactor size than ITER. On the other hand, the major radius (radius of the core curve of torus plasma) of the JA DEMO is ~8.5 m, which is larger than that of ITER: 6.3 m. The plant thermal output of the JA DEMO is ~1.5 GW. These parameters have been selected by considering the design of centre solenoid (CS) coils and the divertor modules. The CS coils are the solenoid coils that are placed at the centre of the torus. In tokamak devices, generation and sustainment of the plasma current is necessary. The CS coils are used to drive a plasma current as an inductive current by changing the magnetic flux that passes through the torus. Although this current drive method is simple and certain, steady-state current sustainment is impossible. Even if pulsed operations are allowed, the achievable current amount and the duration of current sustainment are limited by the specification (the size and the maximum current) of the CS coils. The major radius of the reactor increases with increasing the radius of the CS coils, resulting in increase of the construction costs. Therefore, several methods of non-inductive current drive have been proposed and studied, for example, the use of RF waves or high-energy neutral beams. Most of the past fusion power plant designs assume such a non-inductive current drive method to achieve a steady-state operation with a compact reactor size. In the conceptual design of the Japanese DEMO reactor, however, the CS coil radius has been determined so that the current ramp-up only by the inductive method is accomplished with the technological basis of the superconducting magnet of ITER. The role of the divertor modules is to exhaust impurity particles from the core plasma. The particles are neutralized around the divertor module and pumped away. Therefore, the target plate of the divertor modules is exposed to extremely high heat load. In ITER, this heat load reaches an average of 10 MW m^{-2} plus short periods up to 20 MW m^{-2}, which is larger than the heat load on a spacecraft re-entering Earth's atmosphere. ITER will be equipped with a divertor module with a tungsten target connected to the water-cooled heat sink made of copper alloy, which can accommodate the heat load of 20 MW m^{-2}. Because the fusion output of a fusion power plant is expected to be ~6 times larger than that of ITER, the heat load on the divertor plate also becomes higher. On the other hand, the acceptable heat load of the divertor module is limited to ~5 MW m^{-2} because the copper alloy cannot be used due to the deterioration of its mechanical and thermal properties with the neutron irradiation. To solve this issue, the 'divertor detachment', cooling of the plasma that flows into the divertor plate by an impurity gas puffing, has been proposed as a countermeasure. However, the realization of a stable control of the detachment which is compatible with a good core plasma confinement is not easy. Consequently, the JA DEMO suppresses the fusion output to ~1.5 GW to ensure the soundness of the divertor modules. The latest design values of JA DEMO are shown in table 9.1. For comparison, design values of ITER and Slim CS (past design of tokamak DEMO) are also listed in table 9.1. Figure 9.1 shows the conceptual view of the JA DEMO.

As described above, the technological basis of ITER and LWRs is fully utilized in the design of the JA DEMO. However, there are several issues that require special R&D. One is the development of a large superconducting coil system. In the design of the JA DEMO, the same kind of superconductor and winding method of the coils as those of ITER are assumed. However, both the size and the required conductor current are larger. Therefore, optimization of the coil design including the

Table 9.1. Comparison of primary design values of ITER, JA DEMO and SlimCS.

	ITER	JA DEMO	Slim CS
Major radius (m)	6.2	8.5	5.5
Plasma aspect ratio	3.1	3.5	2.6
Plasma elongation	1.7	1.65	2.0
Plasma current (MA)	15	12.3	16.7
Magnetic field on axis (T)	5.3	5.94	6.0
Maximum field (T)	11.8	13.7	16.4
Fusion power (GW)	500	1420	2950
Fusion gain	~10	17.5	>30
Confinement enhancement, HHy2	1	1.31	1.3
Normalized beta	1.8	3.4	4.3
Normalized density	0.85	1.2	0.98
Bootstrap current fraction	0.85	0.61	0.75

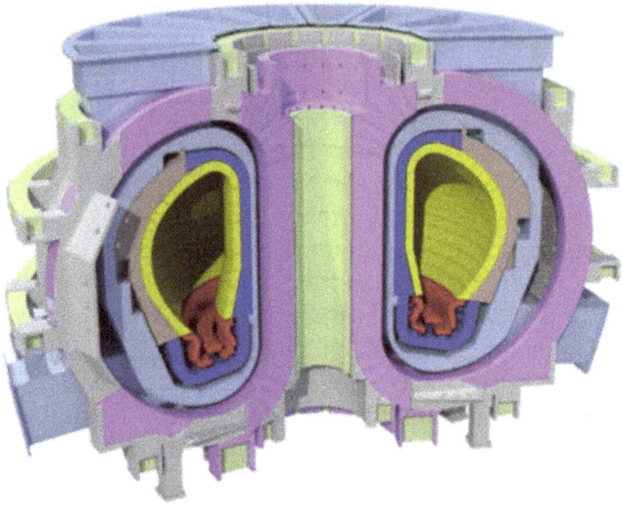

Figure 9.1. Conceptual view of the JA DEMO.

acceptable design and manufacturing errors has been conducted. To realize an electricity generation, a special component that converts the fusion energy (kinetic energy of high-energy particles) into thermal energy, called a 'blanket', needs to be equipped. This blanket module also has the function of producing tritium fuel from lithium-based materials by nuclear reaction between the fusion fast neutrons. This blanket module is a totally new component specific to a fusion reactor and has not been installed in any devices up to now. The JA DEMO selected the water-cooled ceramic breeder (WCCB) concept, which uses pebbles of lithium ceramic as a tritium breeder and pressurized water as a coolant, to utilize the technological basis established in the pressurized water reactor (PWR). The test module of the blanket will be installed into ITER but the operation of this test module is scheduled for a later phase of the ITER D-T operation, which will start around the late 2030s. It is not so easy to achieve sufficient thermal efficiency, tritium breeding performance and the safety feature simultaneously. Then the detailed design optimization has been conducted based on numerical simulations. The manufacturing test of a mock-up module has also been conducted. In the design of the DEMO reactor, the use of several new materials has been considered. For example, the reduced activation ferritic steel will be used for the structural material of the in-vessel components instead of stainless steel to reduce the activation after the operation. Lithium ceramic (lithium titanate) for the tritium breeder and beryllide for the neutron multiplier are also fusion specific materials. In the supporting structure of the superconducting magnet, the application of high strength cryogenic steel which has been developed for the high pressure hydrogen gas tank has been considered. These R&Ds have been conducted with cooperation between the institute (National Institutes for Quantum and Radiological Science and Technology, QST) and several private companies in Japan. The summary of Japan's activity on the DEMO reactor is found in [11].

9.3 Helical reactor design as alternatives

As described in the previous section, the DEMO design activity in Japan has been advanced based on a tokamak-type device. At the same time, several other attractive approaches have been proposed. In Japan, helical and laser devices have been prioritized as an alternative and complimentary approach. In this section, the conceptual design activity and the related R&Ds for a helical device are reviewed.

'Helical' is a collective term for the torus device that generates the magnetic field for the plasma confinement by the external coils only. Thus, the helical devices do not need the plasma current drive and are free from any operational restriction or problem due to the plasma current. This means a steady-state operation is much easier than in tokamak devices. The plant efficiency is also higher than in tokamak devices because the current drive power is not needed. On the other hand, all components including superconducting coils inherently have a complicated three dimensional shape. Therefore, special treatment is needed for the construction and maintenance of the device. The required computational resource for the numerical simulations is also much larger. This is why R&D for tokamak devices are in advance of those of helical devices.

There is a high degree of freedom in the design of external coils of helical devices and various concepts have been proposed. In Japan, the 'Heliotron' system, which consists of two spiral coils and multiple pairs of circular coils, was originally proposed in 1961 [12]. After continuous study with small machines, the Large Helical Device (LHD) was constructed in 1998 and has been operating for over 20 years in the National Institute for Fusion Science (NIFS) [13]. As a kind of helical system, the heliotron system has an inherent advantage in steady-state operation with a high plant efficiency. In addition, the heliotron system has several additional advantages. One is the high design feasibility based on the LHD. Because there are plenty of experimental data and numerical simulation results, reliable core plasma physics designs can be conducted. Compared with other helical systems, the shape of the device is relatively simple and coil winding with a small variation of the curvature and torsion can be realized. The heliotron device also has a robust divertor field structure insensitive to the core plasma condition. It enables flexible design of divertor target in a large space with low radiation environment. There are large apertures between coils and a flexible replacement method of in-vessel components can be adopted.

Based on these advantages, conceptual design activity of the LHD-type helical fusion reactor FFHR has been conducted since 1995 and several reference designs have been proposed. In 2010, the Fusion Engineering Research Project was launched in NIFS and the conceptual design activity and the related R&Ds have been accelerated. In 2011, the conceptual design of a steady-state helical fusion reactor FFHR-d1 was presented [14]. Figure 9.2 shows the conceptual view of FFHR-d1. FFHR-d1 has a capability as a commercial power plant (plant lifetime of >30 years with a fusion power of ~3 GW) with a technological basis that is seen as an extension of ITER technology. However, this design has several issues. One is the

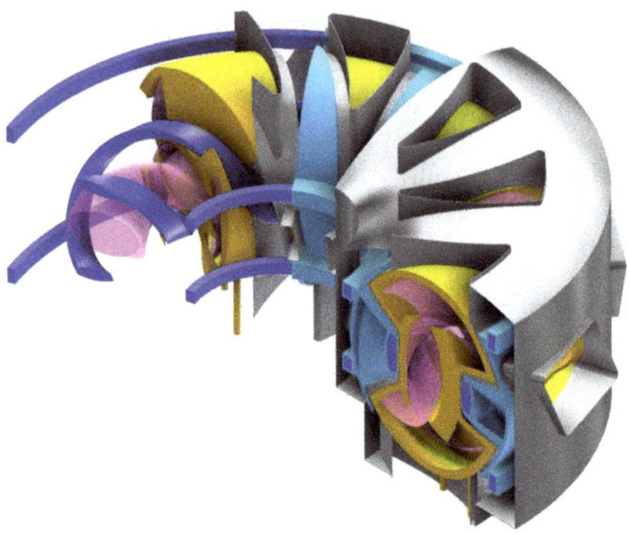

Figure 9.2. Conceptual view of the LHD-type helical reactor FFHR-d1.

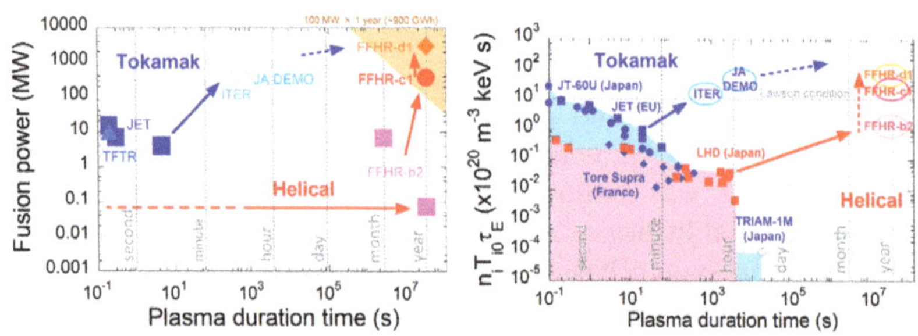

Figure 9.3. Comparison of the achieved/expected performance and strategy to commercialization. The left figure shows the relation between the fusion power and the plasma duration. The right figure shows the relation between the fusion triple product (the product of the plasma ion density, the ion temperature at the plasma centre and the energy confinement time, which is an index of the performance of a fusion device) and the plasma duration.

size of the reactor. FFHR-d1 has a major radius of 15.6 m, which is four times larger than the LHD. This is because the space between the helical coils and the core plasma is quite limited in the helical system. Even with this large reactor size, the minimum thickness of the blanket module is ~80 cm, which is not a sufficient margin. In order to achieve the fusion power of ~3 GW, a certain degree of improvement of core plasma performance from the present achievement in the LHD experiment is required [15].

Thus, step-by-step development strategy towards a fusion power plant with several intermediate devices has been considered [16]. To take advantage of the characteristics of the ease of steady-state operation, a different approach to the tokamak devices has been considered, as shown in figure 9.3. In the current programme based on tokamak devices, the plasma performance (fusion output) of the reactor is increased first. Then the operational period will be increased to the required level as a commercial power plant. In the case of the helical device, the approach that starts with the demonstration of a steady-state operation followed by the improvement of the plasma performance has been considered. The primary design parameters of LHD, FFHR-d1 and FFHR-c1 (intermediate machine between for the demonstration of a steady-state operation with self-sufficient fuel and electricity) are shown in table 9.2.

As an alternative, several innovative engineering design concepts have also been examined. Regarding the superconducting magnet, the use of high temperature superconductors (HTS) is considered to increase the current density and the magnetic field strength [17]. HTS also has a higher thermal stability and cryogenic efficiency compared with the conventional low temperature superconductor. This characteristic can allow larger heat generation in the magnet due to the nuclear heat and resistivity. These characteristics enable a flexible design of the magnet including the neutron shielding and a 'joint winding' of the helical coils becomes possible, which can significantly reduce the construction period of the helical coils.

Regarding the blanket, a self-cooled molten salt breeder is considered [18]. This molten salt is a kind of liquid breeder, which is an advanced concept against the solid breeder with lithium ceramic pebbles. There are three main candidates for the liquid breeder. The other two are liquid lithium and lithium lead. The blanket system with these liquid breeders has a higher thermal efficiency and a simpler structure than those of a solid breeder. Among these three candidates, molten salt has the highest melting point and the lowest heat transfer property. This low heat transfer property will be a disadvantage on the heat removal and thermal usage. However, molten salt has several advantages. It has good tritium breeding performance. The molten salt has low chemical activity and does not react with water. The compatibility with structural material is also good. The operational pressure and the required pumping power are low because of its low mass density ($\sim$2 g cm^{-3}) and low electric conductivity, which leads to low MHD pressure loss.

Regarding the divertor, the use of multiple jets of a liquid metal or multiple flow of a metal pebble as a divertor target has been considered [19, 20]. Such a 'renewable' target can accommodate much higher heat load compared with a solid

Table 9.2. Primary design parameters of LHD-type helical devices.

	LHD	FFHR-c1	FFHR-d1
Major radius (m)	3.9	10.92	15.6
Magnetic field on axis (T)	4[a]	7.3	4.7
Maximum field (T)	9.2[a]	16.4	11.7
Fusion power (MW)		370	3000[b]
Fusion gain		$\sim$10	∞[b]
Helical coil current density (A mm^{-2})	53[a]	55	25
Stored magnetic energy (GJ)	1.69[a]	158	176

[a] Design value.

[b] If confinement improvement is achieved.

target. Efficient pumping through the gap of the liquid metal jets or the metal pebble flows is also expected. The basic R&D for these innovative concepts have also been conducted using the Fusion Engineering facilities in NIFS.

9.4 Issues towards commercialization

As described in the previous sections, conceptual design studies and the related R&D for a fusion DEMO have been intensively advanced in Japan. However, early commercialization of a fusion power plant on the extension of the ITER project is not easy to achieve. The DEMO reactor based on the present physics and technological basis tends to have a large reactor size. The construction period and cost correlate with the reactor size. Of course both the construction period and the cost can be reduced to a certain degree by the optimization of the design and construction method. But the reduction of the reactor size is necessary to significantly reduce the cost and the construction period. However, a compact reactor design is not so easy. The plasma performance strongly depends on the reactor size. Even if the improvement of the plasma performance is achieved, simple reduction of the size with the same plasma performance leads to high heat and particle loads on the in-vessel components. These high heat and particle loads cause difficulties in the design and/or operation of the components and requires frequent replacement of them, resulting in increase of the operational cost. The reduction of the reactor size also leads to insufficient neutron shielding. This insufficient shielding causes shortening of the plant lifetime and increase of the amount of radioactive wastes.

If the fusion power can be reduced simultaneously with the reduction of the reactor size, these issues will be resolved. The heat and particle loads on the in-vessel components are proportional to the fusion power (P) and inversely proportional to the total surface area (S) of the components. The surface area of the in-vessel components is roughly proportional to the square of the reactor size ($S \propto R^2$). Thus, the reduction of reactor size with retaining the same heat and particle loads is possible if the fusion power is reduced while satisfying the condition of $P \propto R^2$. However, this simultaneous reduction of fusion power and reactor size is not necessarily sufficient from the viewpoint of the reduction of the cost of electricity.

In general, the cost of the industrial product is proportional to its mass or volume. Thus, it is expected that the cost of a fusion power plant is proportional to its volume, i.e., cube of the reactor size (R^3). In fact, the components, buildings and facilities in a fusion power plant are categorized into four parts from the viewpoint of the cost dependence. The first is the superconducting coil system. The cost of the superconducting coil system strongly correlates with the total stored magnetic energy, which is roughly proportional to the product of the cube of the reactor size and the square of the magnetic field strength ($\propto R^3B^2$). Second are the other reactor core components (e.g., blanket modules, divertor modules). The cost of this is roughly proportional to the square of the reactor size because these components need to have a certain thickness to fulfil the specific function (e.g., heat removal, tritium fuel production), Third is the system for which size directly correlates with the reactor size (e.g., structural material, reactor building, maintenance facilities).

The cost of this is proportional to the reactor volume, which is proportional to R^3. The other is the system which does not depend on the reactor size (e.g., power supply, plasma heating system, power generation facility). These four parts account for 25%, 40%, 10% and 25% of the total construction cost, respectively. Figure 9.4 shows the dependence of the construction cost on the reactor size. If the reactor size reduces by half, the cost will be 0.4 times smaller in the case where the magnetic field strength is kept constant. However, the fusion power will be 0.25 times smaller in order to satisfy the condition of $P \propto R^2$. Because the electric power is proportional to the fusion power, the cost of electricity will be 1.6 times larger. Actually, reduction of the reactor size while retaining the same magnetic field strength is difficult from the viewpoint of plasma confinement. If the plasma confinement performance is the same, magnetic field strength needs to be changed with the condition of $P \propto R^{-3/4}$. The cost dependence considering this relation is also shown in figure 9.4. In this case, the cost reduced by only 44% and the cost of electricity will be 2.24 times larger.

The adoption of advanced materials or technologies as shown in the previous section can solve the difficulties associated with the reduction of the reactor size. However, establishment of the codes and standards for such advanced materials or concepts is required for the licensing of a power plant. Thus, numerous tests and long-term data acquisition of the advanced materials including the neutron irradiation effect are required. Simultaneous reduction of the reactor size and the fusion output may allow the use of conventional technology. However, improvement of the plasma performance is necessary and the increase in the cost of electricity is inevitable.

In the discussion so far, the commercialization of fusion electricity only is considered. However, a fusion reactor provides not only electricity but also a large amount of heat, high temperature, and monoenergetic fast neutrons in a large area. The large amount of heat can be used for several applications including hydrogen production. The fast neutrons can be used for a component test, isotope production

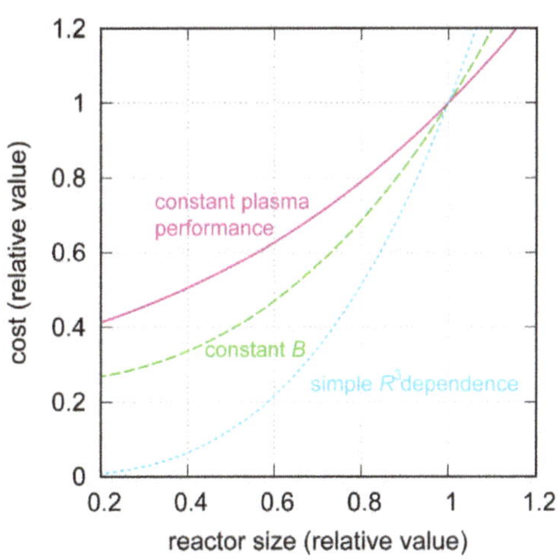

Figure 9.4. Cost dependence on the reactor size.

and the transmutation of fission products. The neutrons can also be used for neutron imaging and medical application if the energy is reduced. These applications can be an early commercialization of the fusion energy.

Finally, I would like to mention the *indirect* effect of fusion R&D. Fusion technology itself has important roles as a leading technology such as space technology. Especially in Japan, the fusion R&D has an aspect of technology security to maintain the technological level and infrastructure of private companies. Fusion energy is expected to be a way to solve the energy and environmental problems of the Earth. But it is more important to leave as many options as possible open for the decision making of future generations. Most of the technologies required for fusion R&D are specific for the fusion reactor. But these technologies have the potential to be the seeds for future innovations. In this regard, steady efforts towards a fusion power plant itself can be a kind of commercialization activity.

References

[1] Joint-Core Team, Report by the joint-core team for the establishment of technology bases required for the development of a demonstration fusion reactor –basic concept of DEMO and structure of technological issues–, NIFS-MEMO-71 (National Institute for Fusion Science, Jan. 22, 2015). (https://mext.go.jp/b_menu/shingi/gijyutu/gijyutu2/056/shiryo/__icsFiles/afieldfile/2015/01/26/1354643_5.pdf)

[2] Joint-Core Team, Report by the joint-core team for the establishment of technology bases required for the development of a fusion DEMO reactor – chart of establishment of technology bases for DEMO, NIFS-MEMO-73 (National Institute for Fusion Science, Mar. 1, 2015). (http://jspf.or.jp/2015/genkeiro/150119_v6.pdf)

[3] Okano K, Kasada R and Ikebe Y *et al* 2018 An action plan of Japan toward development of demo reactor *Fusion Eng. Des.* **136** 183–9

[4] Science and Technology Committee on Fusion Energy, Japan's policy to promote R&D for a fusion DEMO reactor, Dec. 18, 2017. (https://www.mext.go.jp/content/20201012-mxt-000002681_2.pdf)

[5] Kikuchi M 1990 Steady state tokamak reactor based on the bootstrap current *Nucl. Fusion* **30** 265–76

[6] Okano K, Asaoka Y and Hiwatari R *et al* 1998 Study of a compact reversed shear tokamak reactor *Fusion Eng. Des.* **41** 511–7

[7] Kikuchi M, Seki Y and Nakagawa K 2000 The advanced SSTR *Fusion Eng. Des.* **48** 265–70

[8] Nishio S, Tobita K and Ushigusa K *et al* 2002 Conceptual design of tokamak high power reactor (A-SSTR2) *J. Plasma Fusion Res.* **78** 1218–30

[9] Nishio S, Tobita K and Konishi S *et al* Tight aspect ratio tokamak power reactor with superconducting TF coils *Proc. 19th IAEA Fusion Energy Conf.* (Oct. 14–19, 2002, Lyon, France) IAEA-CN-94/FT/P1-21

[10] Tobita K, Nishio S and Sato M *et al* 2007 SlimCS—compact low aspect ratio DEMO reactor with reduced-size central solenoid *Nucl. Fusion* **47** 892–9

[11] Tobita K, Hiwatari R and Sakamoto R *et al* 2019 Japan's efforts to develop the concept of JA DEMO during the past decade *Fusion Sci. Technol.* **75** 372–83

[12] Uo K 1961 The confinement of plasma by the heliotron magnetic field *J. Phys. Soc. Jpn.* **16** 1380–95

[13] Osakabe M, Isobe M and Tanaka M *et al* 2018 Preparation and commissioning for the LHD deuterium experiment *IEEE Trans. Plasma Sci.* **46** 2324–31
[14] Sagara A, Miyazawa J and Tamura H *et al* 2017 Two conceptual designs of helical fusion reactor FFHR-d1A based on ITER technologies and challenging ideas *Nucl. Fusion* **57** 086046
[15] Goto T, Miyazawa J and Sakamoto R *et al* 2017 Development of a real-time simulation tool towards self-consistent scenario of plasma start-up and sustainment on helical fusion reactor FFHR-d1 *Nucl. Fusion* **57** 066011
[16] Miyazawa J, Goto T and Tamura H *et al* 2019 The strategy toward realization of the helical fusion reactor FFHR-c1 *Fusion Eng. Des.* **146** 2233–7
[17] Terazaki Y, Yanagi N and Ito S *et al* 2017 Current-carrying capability of the 100 kA-class HTS STARS conductor for the helical fusion reactor FFHR-d1 *J. Phys. Conf. Ser.* **871** 012099
[18] Yagi J, Sagara A and Watanabe T *et al* 2015 Hydrogen solubility in FLiNaK mixed with titanium powder *Fusion Eng. Des.* **98–99** 1907–10
[19] Miyazawa J, Goto T and Murase T *et al* 2017 Conceptual design of a liquid metal limiter/divertor system for the FFHR-d1 *Fusion Eng. Des.* **125** 227–38
[20] Ohgo T, Goto T and Miyazawa J 2019 A new divertor system using fusible metal pebbles *Plasma Fusion Res* **14** 3405050

Section IV

Challenges and future opportunities

IOP Publishing

Commercialising Fusion Energy
How small businesses are transforming big science
William J Nuttall, Satoshi Konishi, Shutaro Takeda and David Webbe-Wood

Chapter 10

Fusion innovation: understanding the engineering challenges to commercial fusion

Richard J Pearson

10.1 Introduction

Despite slower progress since the turn of the millennium, fusion development is again gathering pace. Publicly funded programmes are supporting ITER (the International Thermonuclear Experimental Reactor) currently under construction in the South of France, to achieve first plasma in 2025. In parallel, over the last decade or so, a plethora of fusion developers backed by private investment are now making progress, with many pursuing non-tokamak reactor concepts. Where ITER's proponents foresee larger, complex, and thus expensive tokamak reactor designs as necessary to achieve commercial fusion, private developers are pursuing less understood, and thus higher risk but potentially faster and cheaper routes. Driven by a belief that the tokamak was down-selected too early (typified by the almost sole focus of globally-coordinated fusion research in support of ITER), private fusion developers are focused on creating revolutionary technologies and demonstrating new physics concepts to satisfy their investors [1]. It is the private fusion industry in particular, with the promise of new pathways to commercialisation that is driving the current zeitgeist with the technology (see chapter 7). This excitement could be considered something of a fusion renaissance. However, I believe that what we are seeing is, in fact, a paradigm shift in the approach to fusion innovation. This shift is from an emphasis on fundamental research for the purpose of advancing understanding of fusion science (with the long-term goal of realising fusion energy), towards one focused more acutely on the engineering challenges that must be overcome to commercialise fusion technology.

'Innovation' is a common term. However, it is often misunderstood. In 2013, technology news publisher 'Wired' quipped that innovation was '*the most important*

doi:10.1088/978-0-7503-2719-0ch10

and overused word in America' [2]. It is, therefore, important to first define what I mean when I use the term in this chapter. Innovation is defined here as *the exploitation of invention*[1]. The majority of government-funded fusion programmes have, up until now, mostly been focused very specifically on technological innovation. Technological innovation can be defined more specifically here as being the exploitation of invention to advance human understanding of science or to achieve technological success. That success is—importantly—not contingent on commercial gain. Predominantly in government-funded fusion programmes, it is expected that any near-term technological success will later be commercialised; but only after that technological success has first been demonstrated. This is evidence of a linear model of innovation which typically underpins the process of innovation across large government-funded science projects, not just in fusion [1, 3]. Such an approach may ultimately result in the successful commercialisation of fusion, but it is not guaranteed to be soon. By contrast, private fusion companies, predicated on the need to get a return on their investment, are operating on a new model which is focused sharply on the commercial exploitation of invention, even if such inventions are not well understood or developed. This is fundamentally different from technological innovation and, as detailed in chapter 7, involves a level of risk that is typically not accepted in—and, understandably, not a main focus of—government-funded scientific research. Moreover, private fusion companies are operating on a different development approach. They are operating on an agile model of innovation in which technology is improved through iterative cycles of building and testing towards a working prototype, which can be scaled up in the same iterative fashion towards a commercial product (see [1]). In consequence, the emergence and early success of these private companies is upending fusion development.

Regardless of the innovation approach, for the ultimate success of fusion there are numerous technical challenges ahead that must be overcome. As noted previously, the majority of current focus in fusion is on the advancement of scientific understanding and demonstration of fusion physics. This is true even in private fusion companies, who tend to be more concerned with creating a working fusion reactor. They see the proof of principle, quite rightly, as the important first step towards commercialisation. As such, while some consideration of the next-step engineering challenges is given, and to a large extent well understood (by both government-funded research programmes and private companies), less emphasis is placed on solving these problems now. If fusion companies are to be commercially successful, they must steadfastly pursue a model of development that considers these engineering challenges at an early stage, in order to begin development of solutions that are both technologically and commercially viable.

Most assessments of the challenges to fusion are, however, focused acutely on the technical aspects. Again, this is evidence of a linear model in which solving the technical challenge comes first, and working to reduce the cost and improve overall commercial viability of the developed technical solution comes later. Such studies are also, typically, focused on specific fusion reactor concepts, mostly for tokamak

---

[1] This definition is taken from former research by the author, see [1, 60].

reactors, and—more specifically still—for reactors based on ITER. If accelerating fusion towards commercialisation is truly the goal, commercial factors should be understood and should, in fact, be used to guide as well as drive technology development. Accordingly, engineering challenges must be viewed through a commercial lens. In my opinion, viewing these engineering challenges from this commercial perspective has been mostly absent in fusion development; although it is inherent in many studies and designs, it is seldom a primary consideration. While fusion has, first and foremost, been seen as a high-tech scientific endeavour, the recent emergence of private fusion companies presents a good opportunity to bring about a shift in how the commercialisation of fusion is viewed when considering the next-step engineering challenges.

The similarities and differences in the next-step engineering challenges across three fusion reactor conceptual spaces—magnetic (MCF), inertial (ICF), and magneto-inertial confinement (MIF)—have been outlined in chapter 2. Irrespective of specific challenges associated with any one conceptual approach, however, there are next-step engineering and technology challenges that must be tackled by all developers pursuing a functioning fusion reactor. As before—and naturally, as the most technologically advanced reactor concept (but not necessarily the best)—most studies, and indeed most R&D, has focused on the tokamak. Consequently, this chapter provides a characterization of these next-step challenges without a focus on any one specific reactor concept. This concept-neutral perspective is intended to act as something of a follow on from the description of the core technical challenges in developing a functional fusion reactor as detailed in chapter 2. However, here I take an opportunity to go further by explicitly considering the commercial drivers associated with these challenges. Looking at both technical and commercial aspects together gives a fundamental understanding of the challenges that stand between where we are now—on the cusp of early-stage fusion reactor demonstration—and where we want to be: ready to launch a commercial fusion reactor. Overall, it is intended to provide those who are unaccustomed to the fusion world—entrepreneurs, investors, and policymakers alike—with a fresh, yet grounded, view on the required steps to achieve commercial fusion.

The chapter is structured as follows. The next-step engineering challenges are separated into five distinct areas, each of which is explored in-depth: fusion materials, tritium breeding, tritium handling, remote handling and waste, and balance of plant systems[2]. The nature of these challenges (and potential solutions) can vary greatly from concept to concept. As such, the broad challenges are first discussed from a concept-neutral perspective. Then, where differences across the three conceptual spaces are apparent, or indeed where they are cross-cutting, these are detailed, with reference to specific real-world development efforts. Subsequently, the engineering challenges are examined from the perspective of commercial drivers—which are, as

[2] This chapter only focuses on approaches to fusion based on the deuterium–tritium (D–T) fuel cycle. Advanced fuel cycles are being pursued only by a minority of programmes. Achieving the physics conditions for p-B11 fusion, for instance, is significantly more challenging than for D–T (see chapter 7). As such, this chapter is focused only on D–T fusion approaches. It is worth noting, though, that while such routes are deemed much more challenging to realise in the near- to medium-term, somewhat ironically, if those near-term challenges can be overcome, then many of the next-step engineering challenges may be much simpler to resolve. Indeed, alternative fuel cycles not based on D–T fusion have significant potential for the long-term.

described, essential for successful innovation. How these drivers should be applied to some of the engineering challenges outlined earlier in the chapter, with reference to several interesting examples, is then discussed. Finally, an outline of an innovation strategy for commercially driven fusion technology development, regarding the approach needed by both private developers and public fusion programmes, and the collaboration between both, is considered.

10.2 Engineering challenges to commercial fusion

This section details the challenges that are more broadly shared across the three fusion reactor conceptual spaces (MCF, ICF and MIF). While specific designs vary, even within conceptual spaces, challenges in the following five areas must be resolved to realise commercial fusion:

1. Fusion reactor materials
2. Blankets for tritium breeding and power generation
3. Tritium handling and management
4. Waste and remote handling
5. Balance of plant systems

10.2.1 Fusion reactor materials

The development of suitable materials underpins almost of the challenges that have been explored throughout this book, and those challenges which will be explored later in this chapter. For instance, fuel targets for ICF reactors and divertors for tokamak reactors are both fundamentally materials issues (see chapter 2). However, they are also very different problems. Different reactor concepts will be subject to a wide variation in the levels of electromagnetic, neutron and particle radiation. Furthermore, there are different materials requirements across the various systems within even the same type of fusion reactor, let alone across concepts from different conceptual spaces. Despite the diversity, scale, and complexity of the challenges associated with developing fusion materials, however, it also represents perhaps the most cross-cutting issue.

In addition to advanced materials development for specialist components[3], and specific materials for tritium breeding blankets (which are detailed in the next section of this chapter), perhaps the most important challenge of all is to develop functional materials to be used inside a fusion reactor chamber. So-called plasma-facing materials receive the highest thermal, radiation, and mechanical loads. To understand this further, it is important to understand that the neutrons impacting the plasma-facing materials in a D–T fusion reactor have a peak energy of 14.1 MeV

[3] The precise manufacture of, for example ICF targets (or MIF fuel targets, in the case of MagLIF); the development of HTS tape for high-performance magnets in MCF reactors; and the advancement of materials-based technologies such as lasers or acoustic drivers (for ICF and MIF reactors), are all fundamentally materials challenges. However, as these are concept-specific materials engineering challenges, they are not covered in detail here. Refer instead to chapter 2 and references therein.

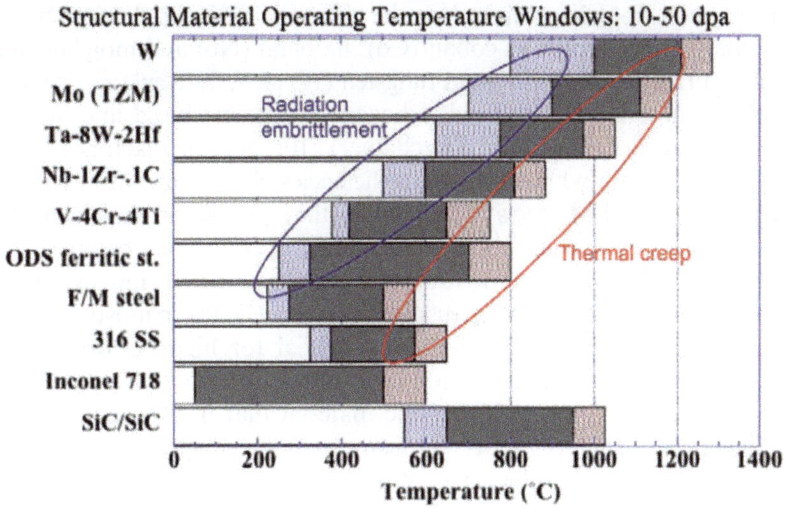

Figure 10.1. Operating temperature window for structural materials for fusion. Reproduced from [6], copyright (2000), with permission from Elsevier.

and the bombarding neutrons have an energy in the multi-MeV range. As such, the effects of radiation damage in a D–T fusion reactor are expected to be much greater than for fission reactors [4]. Novel structural materials are being developed to handle the harsh nuclear environment in fusion reactors. However, developing such materials is challenging. Many materials, metals in particular, due to embrittlement under neutron irradiation and thermal creep, have a limited window of allowable operating temperatures [4–6]. At certain temperatures and levels of neutron irradiation[4], structural materials face radiation-induced hardening and embrittlement (including helium embrittlement), phase instabilities due to precipitation, creep, and swelling (due to void formation) [7, 8]. Figure 10.1 shows the operating range of several candidate fusion materials, which illustrates two things. Firstly, the operating temperature of any one material is limited by the effects of thermal creep at higher temperatures and radiation embrittlement at lower temperatures. Secondly, using various materials in combination with one another is not always possible, as those materials may have entirely different operating windows[5].

Certain grades of steel that are widely used in nuclear fission and other high-temperature applications include alloying elements that, when irradiated, become activated. A particular grade of steel that is commonly known as RAFM steel (Reduced Activation Ferritic Martensitic) has been developed to be functional under

[4] Radiation damage is typically measured by displacements per atom (dpa). Neutrons produced by certain nuclear reactions, including D–T fusion, have sufficient kinetic energy to dislodge atoms in the lattice of materials that they encounter. The radiation damage caused by these collisions is described by the dpa, where a dpa of 1 corresponds to every atom in a material lattice having been displaced from its original site once (see [7, 8, 11] for a more detailed description).

[5] In addition to the reactor vessel itself, the operating temperature window issue also particularly affects the design and performance of the tritium breeding blanket.

neutron irradiation. RAFM steels reduce the effects of radiation damage by replacing typical alloying elements such as cobalt (Co), niobium (Nb) and molybdenum (Mo), with tantalum (Ta), vanadium (V) and tungsten (W) [4, 7, 9]. Replacing solutes in this way reduces the amount of long-lived radioactive waste produced in a fusion reactor (discussed in more detail later in this chapter) [10]. Fusion reactors must operate well above 600 °C to attain higher thermal efficiencies. Such temperatures are currently not possible using RAFM steels. Thus, while they provide a significant reduction in neutron-induced activation, they have a limited operating temperature of around 500 °C–550 °C. More advanced structural materials are being explored as an alternative to RAFM steels. Oxide dispersion steels (ODS), nanostructured ferritic alloys, and vanadium alloys all offer the potential for high performance at high temperatures and under neutron irradiation. Similarly, silicon carbide composites (SiC/SiC) are being explored as a ceramic material that has the potential for high thermodynamic performance under the same extreme nuclear conditions [7, 11]. However, SiC/SiC as well as the array of advanced metal alloys, both of which are necessary materials for different functions, have not yet been sufficiently developed and further R&D is needed [12–14].

In addition to structural materials for reactor vessels that provide the high vacuum environment required to maintain a high-quality plasma and to confine tritium [15], materials that can withstand physical contact with the steady-state plasma are required for MCF reactors. Plasma-facing materials must withstand high thermal and neutron loading while avoiding erosion, whereby the plasma physically contacts and degrades the materials on plasma-facing surfaces which can contaminate the plasma. The loading is particularly extreme in the divertor region, where the plasma is in constant contact with the material surface and where erosion by high-energy particles from the plasma is particularly high [11]. Typically, despite being a heavy element and thus having the potential to pollute the plasma, tungsten is selected as the most suitable plasma-facing material for its low erosion, minimal neutron activation, and limited tritium retention [8, 11, 16]. Tritium retention in materials is another significant materials challenge, and which can also have an impact on plasma performance. This is because if tritium becomes trapped within the materials in the reactor chamber, it cannot then contribute to the fuel cycle. Materials must be chosen judiciously, with those that avoid the substantial retention of tritium favoured. Despite having useful thermal and plasma performance, carbon—which was used in some of the experimental programmes on the JET tokamak—is a good example of an unfavourable material as it has a particularly high tritium retention rate, despite it having other beneficial properties. There is also a safety aspect to tritium retention, too, as tritiated components are harder to handle when they are removed from the reactor during maintenance at the end of life. Related issues with tritium management are detailed later in this chapter.

An ICF reactor must handle radiation loads as well as the pressure wave that arises from the implosion of the fuel target. In NIF, an ICF reactor (see chapter 2), 45 MJ of energy is produced in a single pulse. This is equivalent to exploding about 11 kg of TNT (tri-nitro toluene, an explosive chemical compound) [11]. In a commercial ICF reactor, the first wall must be capable of withstanding this explosive

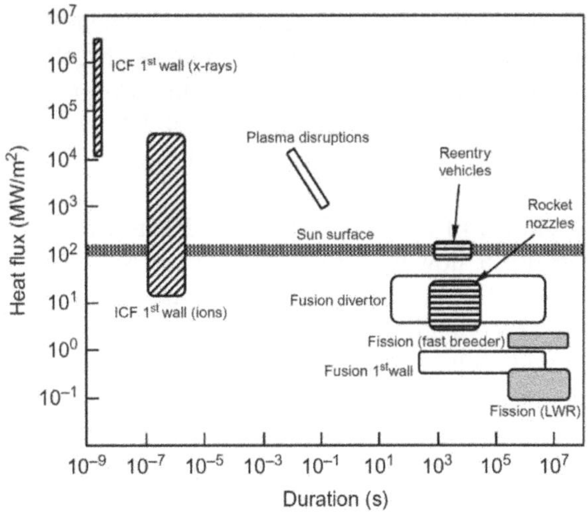

Figure 10.2. Power loads and the duration over which they occur on materials in various nuclear and extreme environments. Reproduced from [17], copyright 2019, with permission from Elsevier.

force, repeatedly. In addition to this force, the damage from high-energy fusion neutrons, and the high-speed shrapnel from the destruction of the fuel target, it must also withstand ultra-high and transient heat flux from ions and x-rays, as shown in figure 10.2. In MCF reactors, such a heat load is not experienced even in the event of a plasma disruption. ICF systems must tolerate these intense blasts from the pulsed fusion reaction, which will deposit enormous quantities of heat over very short time intervals. In a commercial reactor, this is likely to be continuous, perhaps for a duration in the order of years. Such cyclical loading will result in thermomechanical fatigue [17]. Indeed, understanding the harshness of such an environment led to the realisation that liquid metal walls may be a necessity for ICF reactors[6]. Whilst some solid-walled structural materials are still being considered for ICF reactors, 'self-healing' liquid metal walls have been explored as a potentially advanced solution [7, 18, 19]. Liquid plasma-facing materials are effectively immune to neutron damage, hence 'self-healing'. They would also be more tolerant of high flux heat loads (including for transient operation in pulsed ICF and MIF reactors) [20]. A liquid first wall, acting as a barrier between the exploding target and the first wall of an ICF reactor, could protect the reactor structure from x-rays, ions, and neutron radiation damage [21]. It may also have the added benefit of decreasing the amount of radioactive waste that is produced by effectively shielding the solid structural materials of the reactor vessel. Indeed, while also a key component in the fusion reaction mechanism (for compression, see chapter 2), some MIF reactors that employ metal liners are, in fact, predicated on the idea that the liner will protect the reactor vessel from any debris and the shockwave produced in the fusion reaction [22].

[6] Other issues relating to the ICF laser repetition rate must be considered when determining the viability of liquid versus solid walls for ICF reactors, as detailed in [19].

For example, General Fusion's concept (see chapter 7) intends to use a liquid metal liner of lithium-lead, where the liner has the purpose of absorbing the impact of the neutrons from the fusion reaction. Further, this configuration would simultaneously breed tritium (due to the lithium in the liquid metal liner, discussed in the next section of this chapter), as well as transfer the energy from the neutrons as heat, potentially simplifying the overall design [23].

Another novel example of a liquid first wall is First Light Fusion's ICF reactor concept, which will use a flowing liquid 'curtain' surrounding the fusion reactor vessel which will absorb the shockwave as well as the debris from the implosion of the fusion fuel target. Flowing liquid metal divertors are also now being considered for tokamaks to provide self-healing of the plasma strike surface [24][7]. Such liquid metal plasma facing materials show substantial potential to increase the lifetime of the plant, reduce the failure rate of in-vessel components, and would also allow active (online) maintenance. However, the thermal hydraulics, pumping efficiency, tritium and impurity removal, and heat extraction from liquid metals used inside a fusion reactor environment remain as key gaps [17]. Additionally, due to the high vacuum required as well as the presence of a large magnetic field, there are potentially much greater challenges associated with the hydrodynamic control of flowing liquid metal inside an MCF reactor.

ITER will provide testing of materials in a reactor-relevant environment, but only up to a damage threshold of 3 dpa [16]. Experiments conducted on materials in proxy environments can push the damage level beyond 3 dpa, meaning that ITER will not significantly advance the boundaries to facilitate advancements in fusion materials. Moreover, it cannot be used to validate any of the advanced fusion materials being employed [4]. Regardless of reactor concept, therefore, a dedicated materials facility for testing—such as the widely suggested idea for an 'IFMIF' (International Fusion Materials Irradiation Facility)—would usefully provide a simulated fusion-relevant environment [16, 17]. Such a facility must provide a long duration testing at high heat fluxes, high magnetic fields, and with simulated force of ICF or MIF shockwaves etc. It must be developed as soon as possible to be in parallel with reactor development efforts, if commercially viable materials solutions are to be developed for any and all reactor concepts in the order of years, rather than decades.

10.2.2 Blankets for tritium breeding and power generation

Tritium, one of two fuels required for D–T fusion reactors, is available only in limited quantities. It is sourced almost entirely from CANDU-type fission reactors. However, tritium sourced from CANDU reactors is wholly insufficient to support a commercial fusion programme, as the production from all CANDU reactors on Earth is around 2–3 kg per year, while a 1 GW (thermal) commercial fusion reactor will require around 56 kg per year [25, 26]. As such, for a commercial fusion programme, it must be considered that tritium supply from external sources is

[7] Liquid breeding blankets are detailed in the next subsection.

effectively zero. Instead, tritium must be produced *in situ* in a fusion reactor via tritium breeding. Tritium is created when a neutron from the D–T fusion reaction interacts with lithium in a tritium breeding 'blanket' surrounding the reactor. The blanket has two primary purposes. Firstly—as is obvious—it is used to breed tritium for the reactor. Secondly, it acts as the medium to transfer the kinetic energy of the fusion neutrons as heat. Accordingly, the blanket is a crucial interfacing system for the reactor core with the Balance of Plant (BoP) systems, which together allow useable energy to be produced from the fusion reactor (detailed later in this chapter). As alluded to earlier, the potential for using liquid metal in ICF and MIF reactors (in the form of a liner for MIF), which can act as both the first wall and as the blanket, means that blanket configurations across the three reactor spaces vary widely. Despite this, the fundamental principles, materials, and even some aspects of the designs—including, for example, the mechanisms to extract tritium from the blanket—are quite similar.

Before we can delve into specific technical challenges, however, it is first important to understand the key performance parameter for tritium breeding blankets: the tritium breeding ratio (TBR). TBR is a measure of the tritium produced in the breeding blanket against the tritium consumed in the fusion reaction (or otherwise lost, as detailed subsequently). All fusion reactors must achieve a TBR of above unity to be tritium self-sufficient, whereby the fusion reactor can sustain itself using lithium for breeding, rather than an external source of tritium. Although a TBR of precisely 1 may appear to be sufficient for a commercial fusion reactor, in reality, the TBR is typically required to be approximately 1.10–1.15. This margin is necessary to account for [27]:

- The radioactive decay of tritium (at a rate of 5.47% per year).
- System losses due to, for example, the retention of tritium in materials (see the previous section of this chapter).
- The time taken to recycle tritium through the tritium systems (see the subsequent section of this chapter).
- The build-up of a reserve inventory for reactor restart or the start-up of other reactors.

Additionally, breeding blankets cannot cover the entire of the plasma-facing surface for all reactor designs. Furthermore, even if they can cover a large portion of the surface area, not all neutrons from the fusion reactor will successfully interact with a lithium atom. While the issue of blanket coverage is well characterized and accounted for, especially in MCF reactor designs, it will greatly impact the design of any commercial fusion reactor. All reactors, for example, require port space for diagnostics and fuelling systems. MCF reactors require space for the divertor, and MIF and ICF reactors require ports for injectors and lasers, respectively [27]. This lost space for breeding means that, typically, special materials that can increase the neutron yield—so-called neutron multipliers—must be used alongside lithium in the blanket. Beryllium and lead (as well as other specific materials, including uranium; typically avoided as an element associated with nuclear fission reactors and for its links to nuclear weapons), when struck by a neutron at a specific energy, undergo a fission reaction to produce two neutrons [4, 28, 29]. Beryllium can be incorporated in solid form into pebble bed blankets but also in liquid form in molten salt blankets, while lead can be combined with lithium to form a liquid metal eutectic. The inclusion of multipliers serves to increase the neutron yield to make up for the aforementioned losses[8].

Blanket coverage in an ICF or MIF reactor has the potential to be much higher than for an MCF reactor such as a tokamak which—as shall become clear—may lead to a more efficient commercial reactor design. Principally, this is because tokamaks require at least one divertor, and need a solid first wall in front of the blanket system since employing a liquid plasma-facing material is extremely challenging under the high vacuum environment and high magnetic fields.

Blankets must be designed around achieving sufficient TBR as the key performance goal, but they must also be developed to employ functional materials that can withstand the high-energy neutron environment, to efficiently extract the tritium that has been generated, and to effectively transfer the neutron heat for power generation[9]. Three main types of breeding blanket are currently being investigated: solid beryllium-based, liquid lead-based, and beryllium-based molten salt. As lead-based and molten salt blankets use a flowing liquid medium, they can be 'self-cooling', where the flow of the liquid medium carries the heat without the need to use an external coolant. Such designs can also be cooled by an external coolant, such as helium or water, but solid breeding blankets, on the other hand, can only be cooled by way of an external coolant. As detailed in the previous section, the choice of blanket breeding medium and coolant is fundamentally a materials challenge, mostly because of allowable operating temperatures and neutron damage limits. All blankets can employ water as a coolant, but this may limit efficiency due to low maximum operating temperatures [30]. A schematic of a solid beryllium-based blanket for a tokamak-type reactor, which is used here to illustrate the general structure of a blanket, is provided in figure 10.3.

All breeding blankets are subjected to high-energy neutrons, elevated temperatures (including thermal transients, even in MCF reactors operating at steady-state), and large magnetic fields and shockwaves in the case of MCF, and MIF or ICF reactors, respectively. Structural steels and advanced ceramics, such as SiC/SiC (which are particularly important for very high-temperature and self-cooled blanket concepts), will be necessary if high thermal conversion efficiencies are to be achieved and if radiation-induced effects, such as swelling, are to be avoided. They are also crucial in MCF reactors, in particular, to isolate the electrically conductive flowing lithium-lead metal from the surrounding steel structure of the blanket, to avoid magnetohydrodynamic (MHD) effects [28, 32]. However, SiC/SiC, as well as other

[8] It is also important to note that lithium-6, the isotope that makes up only 7.4% of natural lithium on Earth, has the higher probability to produce tritium than lithium-7. As such, lithium must be enriched in the lithium-6 isotope, which presents some technical but also intrinsic commercial challenges [63]. An optimal solution would be through use of a breeding blanket that employs natural lithium [60].

[9] Due to neutron interactions in blanket materials, blankets have an energy multiplication effect, which ranges from 1.05 to 1.25 dependent on the design, chosen materials, and structure. See [21].

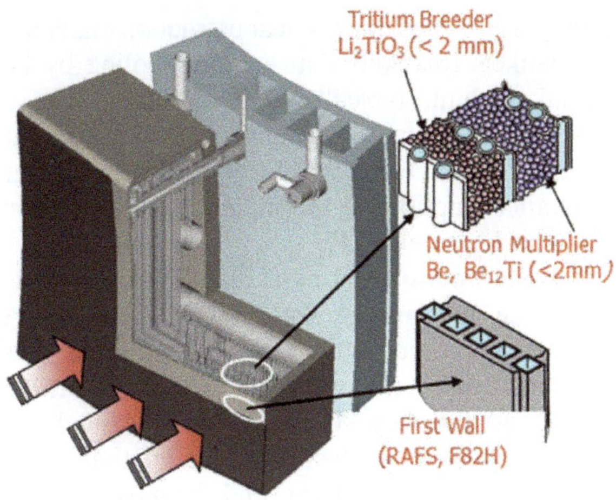

Figure 10.3. Cross-section of a solid-type beryllium blanket (for a tokamak-type reactor). Reproduced from [31] with permission (©IAEA).

advanced materials, require further development and testing to validate their use [12, 14]. Solid beryllium-based blankets do not suffer the same MHD effects as some flowing liquid metal blankets, but they do have an altogether different, but similarly limiting problem relating to the maximum allowable operating temperature. Current designs are limited to the maximum operating temperature of RAFM steels (as detailed earlier, this is around 550 °C). To operate at a higher temperature would most likely require the development of creep-resistant high-temperature metals for the blanket structure [4, 28]. However, such concepts are also limited by the thermomechanical properties of the solid beryllium breeder material, as well. While molten salt blankets do not suffer MHD effects and are capable of higher temperature operation—thus potentially avoiding the key drawbacks of both liquid and solid blankets—they face several other, quite unique problems. The high viscosity of molten salts means that high-power pumping is required. As with lead-lithium blankets, to realise the potential for very high-temperature operation, advanced materials must be developed [32]. It is uncertain, for example, whether SiC/SiC is compatible with molten salts. There is an added need to develop any materials to be corrosion resistant, as tritium fluoride is produced when tritium and fluorine (F) in FLiBe and FLiNaBe salts interact (these salts are the two most common under consideration for fusion breeding blankets) [33].

Tritium extraction—the recovery of tritium from the breeding medium—is a challenge shared across all blanket types and reactor configurations. In solid breeders, it is highly temperature-specific. In liquid and molten salt breeders, it depends on the efficiency of removal from the flowing liquid [32]. Similarly, heat extraction is a challenge shared across blanket types and reactor configurations. It is particularly challenging in solid breeders cooled by helium, as intense nuclear heating causes large temperature gradients which results in unacceptably high pressure drops across the helium coolant channels [28, 34]. Similarly, low thermal conductivity in molten salt blankets, while reducing heat losses, makes heat extraction challenging [32, 33]. Dual-cooled lithium–lead blankets (blankets that employ cooling by both flowing liquid metal and an external coolant, typically helium) might offer a solution to the challenges of tritium and heat extraction, but face challenges relating to, inter alia, the aforementioned issue of MHD effects [35, 36]. Self-cooled blankets may offer the simplest yet most advanced option, if the materials required for such blankets—particularly SiC/SiC—can be developed [13].

Blankets have long been considered as a reactor component that will require frequent replacement due to damage and lithium burn-up. Accordingly, it is regarded as a component that dictates the lifetime of, and the level of maintenance required for a commercial fusion reactor. As mentioned earlier, ICF and MIF could avoid some of the challenges faced by MCF reactors by using flowing liquid metal or molten salt blankets to act as a self-healing material that can function simultaneously as the first wall, breeding medium, and primary heat transfer fluid. Such breeding blankets would also allow for online control of lithium concentration, i.e. the level of lithium-6 enrichment, to actively control the TBR [37]. Liquid blanket designs have been developed for ICF reactors that use either 'thin' or 'thick' liquid walls (in addition to some designs with solid walls), as shown in figure 10.4. Such blanket designs must still allow for penetrations of the ICF fuel target injector and laser beam(s). It must also remain stable and uniform throughout the pulsed operation of the reactor. As noted previously, First Light Fusion is designing a liquid 'curtain' (rather than a blanket) made of pure lithium, to be used for tritium breeding and heat extraction, as well as protection for the surrounding reactor structure. One challenge with this design, however, is that the high solubility of hydrogen isotopes in lithium results in low permeation rates[10]. This might make extracting tritium difficult [38]. Similarly, some MIF concepts envisage liners made of liquid metal as having the dual function of acting as a medium for the fusion reaction itself, as well as for tritium breeding. Tin–lithium, lead–lithium, and pure lithium are being explored as potential liner materials, but the development of tritium extraction systems for these flowing metals is identified as a significant technical gap [20, 39].

While solid and liquid breeding blanket designs are being developed for testing in ITER, there are currently no published plans for developing or testing ICF and MIF blanket concepts. As such, like several other technology challenges, MCF blanket designs are the most advanced—that does not necessarily mean that current designs are optimal. Due to the potential for greater blanket coverage, the possibility to avoid high levels of lithium enrichment (due to the potential for greater blanket coverage), MHD effects, and the dual function of acting as a self-healing first wall, non-MCF approaches may have an advantage. In summary, regarding the design of tritium breeding systems, despite being at lower technology readiness compared with

[10] Such low permeation can also be an advantage, as tritium is less likely to permeate from the blanket into the surrounding reactor structure. Despite that, this effect, known as tritium migration, is a key problem. Tritium that permeates from the blanket into the cooling loops, for instance, can cause the heat exchanger to become tritiated, contaminating the BoP systems (detailed later in this chapter).

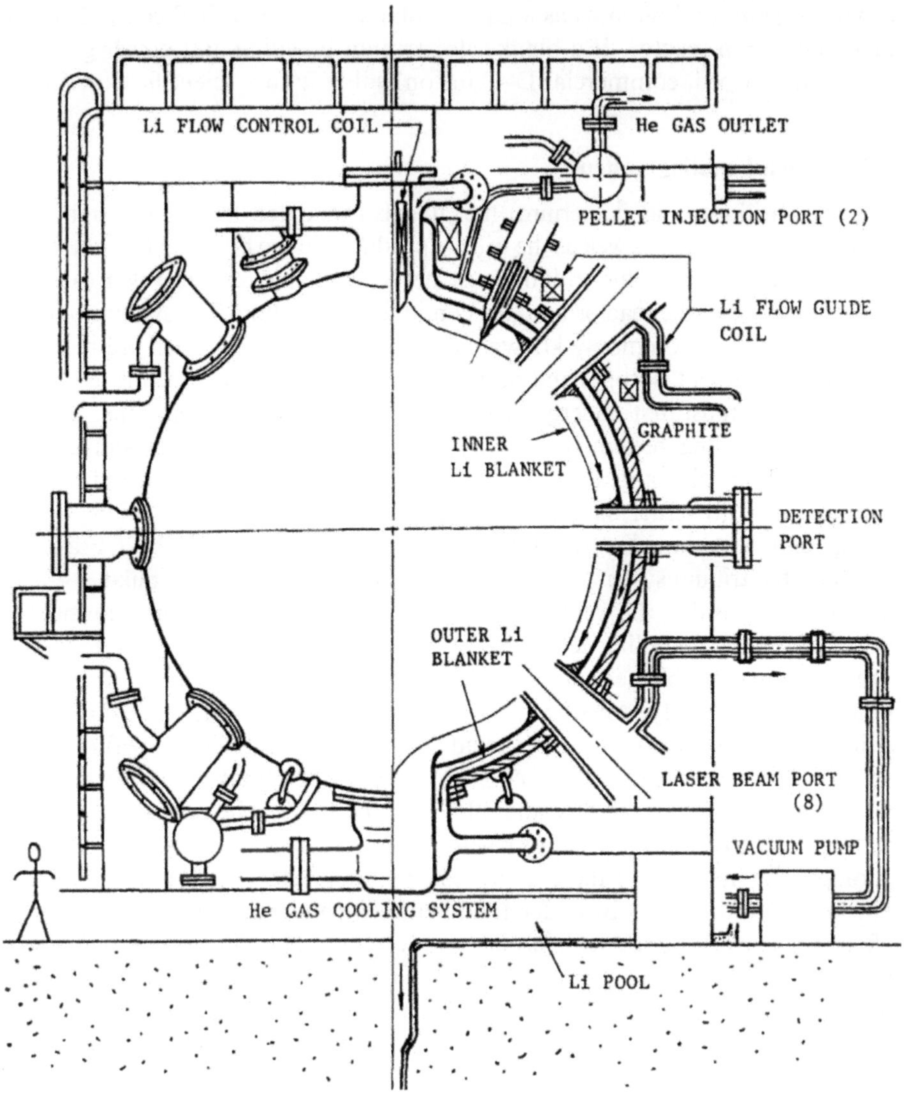

Figure 10.4. Liquid lithium-walls in the 1980s SENRI-I conceptual ICF reactor. Reproduced from [18]. ©Atomic Energy Society of Japan. Reprinted by permission of Informa UK Limited, trading as Taylor & Francis Group, www.tandfonline.com on behalf of Atomic Energy Society of Japan.

MCF reactors, if solutions to the engineering challenges can be developed then ICF and MIF reactors may present a better pathway to a commercial fusion reactor.

An important final point to be made is that as with fusion materials, a facility to test the tritium breeding blanket concepts does not yet exist. All concepts, including for MCF reactors, despite decades of dedicated R&D, are at low technology readiness due to the lack of such a facility (or, indeed, an operating fusion reactor). Novel and commercially viable designs must be developed in conjunction with a means to test and validate them in a fusion relevant environment. This need is somewhat urgent, and remains as a gap, as blankets are a critical component for a commercial fusion reactor. If a blanket for tritium breeding and power generation cannot be developed, commercial D–T fusion will remain a scientific experiment[11].

10.2.3 Tritium handling systems

The unspent tritium (and deuterium) fuel that is evacuated from the fusion reactor chamber via the exhaust, as well as the tritium produced in the breeding blanket, must be processed and recycled. This tritium must be accumulated and recycled to be later introduced back into the reactor to allow continuous operation. The systems to control and manage this are commonly known as the tritium systems, which are responsible for separating different isotopes and purifying them into different streams to be reused as fuels. Tritium must first undergo separation and purification from other gases and compounds, including helium, carbon, and oxygen. These effluents and gases are detritiated and then discharged to the environment [8, 37, 38]. The tritium and deuterium isotopes required to refuel the reactor are then combined with the tritium from the breeding blanket. Although exhaust and fuelling systems across MCF, ICF and MIF spaces vary, the tritium systems for each reactor concept are largely similar.

Before we consider the minor differences, however, let us first consider the function of tritium handling technology and systems. The technology for isotope separation is well-established, in particular, due to the experience of tritium removal facilities for CANDU-type fission reactors [37]. Tritium systems for fusion reactors have also been developed and experimentally validated in fusion experiments, too, most notably in the tritium systems commissioned for tokamaks that operated in the 1980s and 1990s (JET in the UK and TFTR in the USA) The design for ITER has built on these experimental validations and aims to further advance tritium system technology. A schematic of the tritium systems for ITER is shown in figure 10.5.

The key point that must be understood before we move on is that any component in a fusion reactor that produces, handles, or even interacts with tritium—in any way—must interface with the tritium systems. Most obviously, the tritium systems must, therefore, interface with the main reactor chamber and the blanket. The tritium contained within all such systems and components makes up the overall tritium inventory of the reactor. As such, the tritium systems have a key function for tritium accountancy, as well as for processing. Additionally, the tritium systems connect to systems entirely external to the reactor vessel to, for example, monitor the composition of the atmosphere in the reactor building. The tritium systems thus act as an important confinement barrier to avoid tritium contamination [38].

The tritium systems connect the exhaust to a range of cryogenic installations which separate the different hydrogen isotopes and compound species, due to the different boiling points of the gases. Tritiated water is handled separately by a water detritiation system (in reactors that use water for cooling). The tritium breeding blanket is also connected to these systems—whether cooled via helium or water—and the tritium

[11] With thanks for Aniceto Goraieb of Karlsruhe Beryllium Handling Facility, and Chris Dorn of Beryllium 4 Fusion, for this characterisation.

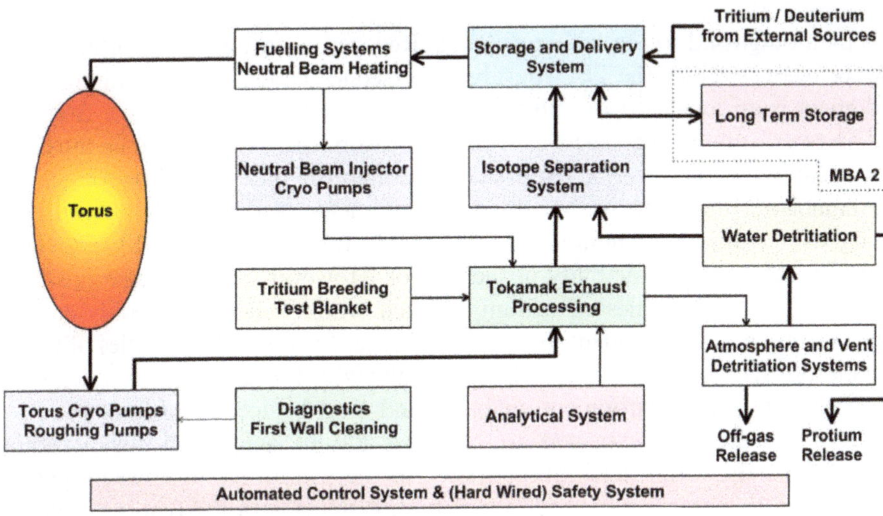

Figure 10.5. Tritium fuel cycle systems for ITER. Reproduced from [37], copyright (2008), with permission from Elsevier.

produced from the breeding medium extracted. All separated isotopes are then redirected to the fuelling systems—this is the pellet factory, in the case of an ICF reactor—and then evacuated to the atmosphere and vent detritiation systems, or sent to isotope storage, where tritium is stored in zirconium cobalt or depleted uranium getter beds that contain it as a gas in their material lattice. The tritium can then be released by elevating the temperature of the bed when it is needed [8].

The tritium systems are critical for reactor performance, safety, and the environment. As detailed in the previous section, the quantity of tritium available for reactor operations depends on the TBR. A key factor in determining TBR is the time taken for the tritium systems to process tritium from the exhaust and blanket to be returned to the reactor as fuel. Tritium must be processed quickly enough—which also necessitates that significant quantities must not be retained in materials—to enable the reactor to operate with a relatively small inventory, perhaps in the order of a few kilograms of tritium. The performance of the tritium systems is thus a key driver for the performance of a commercial reactor [28, 40].

As detailed in the previous sections of this chapter, because tritium is highly permeating it can become embedded, or 'retained', in materials. The retention of tritium is higher in some materials, for example carbon. The principal issue here is that tritium trapped in materials is not available to be used as fuel. Any tritium retained in materials must, therefore, be removed. So-called tritiated components can be detritiated in one of two ways: via removal *in situ* or in a hot cell. For MCF reactors, *in situ* tritium removal could be achieved via high-temperature baking in an oxygen atmosphere using plasma discharges, which physically removes embedded tritium from materials [41]. Hot cells, which are detailed as part of the remote handling systems detailed in the next subsection, are a separate area away from the fusion reactor that handles tritiated—as well as radiation-damaged—components, treating them before disposal. Tritium released from tritiated components in the hot cells is then outgassed to the tritium systems. Recovering tritium is important; as if it cannot be recovered then the TBR must be increased to account for lost tritium, but the added effect is an increased overall tritium inventory. This is undesirable from both safety and performance perspectives. For this reason, the tritium systems must have high availability as they will be required to continuously process tritium—even when the remainder of the plant is shut down for maintenance [38]. Both the process of 'purging' tritium *in situ* and removal of components to be detritiated in the hot cells has a direct impact on the availability of the reactor. However, both are also critical for safety. Such large quantities of tritium retained within reactor materials could be released during a severe accident or a vessel leakage. Even small quantities of tritium, in certain forms, can be hazardous to human health [42]. Tritium from all systems must not be released to the environment unabated and must be accounted for.

Tritium systems are expected to be cross-cutting across all fusion reactor concepts, from the water detritiation system to the tritium storage system [37, 38]. In addition to the interface between the tritium systems with the fuelling systems—where, for an MCF or MIF reactors it is turned into frozen fuel pellets or injected as a gas, for ICF reactors (and some MIF reactor concepts) it is fabricated as a part of a fuel capsule. A key difference, however, is in the connections between the tritium breeding blanket and the tritium systems. As detailed in the previous section of this chapter, extracting tritium from breeding blankets—whether solid or liquid-based breeders are employed—remains a technical challenge. While MCF reactors employ a standalone blanket system, regardless of whether a solid or liquid-based breeder is employed, it is positioned behind a separate first wall. For ICF and MIF reactors, however, a liquid wall that acts as both the plasma-facing first wall and the breeding blanket may be employed. Evidently, therefore, how the tritium systems interface with the blanket systems will vary including across concepts within a single conceptual space. For ICF reactors, in particular, the debris from the target implosion must be extracted and processed by the tritium systems, in addition to fusion products, unburnt fuel, and impurities, to be recycled to the target factory (or manufacturing system) [21, 38]. This is a unique challenge. However, there may also be benefits. Losses due to tritium retention in materials may be largely avoided as the liquid tritium breeding blanket also acts as the first wall thus allowing the tritium to be passed directly to the tritium systems [38].

Finally, while tritium systems are at a reasonably high level of technological readiness and designs for commercial fusion reactors have been conceptualized (see [37, 38]), developing tritium systems suitable for a commercial reactor is not a simple extrapolation from the current state-of-the-art. ITER will provide operational experience and validation of more advanced tritium systems, including with larger quantities of tritium (in the order of kilograms) and at higher circulation (fuelling and exhaust) rates than has been seen in previous fusion experiments. Whether this tritium fuel can be supplied continuously—which is a crucial requirement for a commercial reactor—remains as an experimental aim of the project. However, the integration of the tritium systems on ITER is a unique challenge—one that has never been carried out before at that scale [37]. A commercial reactor will require both low residency times for tritium in the reactor (fast processing speeds), but also must handle large quantities of tritium from the breeding blanket to produce a specific amount of tritium. If the breeding blankets produce too much or too little tritium (due to a TBR that is too high or low), or too much tritium is retained within the tritium systems or reactor materials, it will impact the tritium balance of the reactor and may give rise to safety or security concerns [37]. As this problem is related directly to the demonstrating of tritium breeding blankets, it thus remains an open technical gap as none of the designs has been validated through experimentation.

10.2.4 Waste management and remote handling

As is clear from the earlier description of the materials challenges, D–T fusion reactors will produce radioactive waste. While the challenge of waste management is a well characterized and known issue, however, it is not widely advertised, particularly in public communications which tend to promote fusion as a source of energy that does not produce any long-lived waste[12]. First, it is important to understand that radioactive waste from fusion will not be the same as that from fission reactors—most importantly, there will be no fissile nuclear waste. Waste from fusion reactors will come in two forms: materials that become radioactive because of neutron activation and components that have become tritiated. As with the tritium systems, the handling of radioactive waste from fusion will be a challenge that is shared across the fusion conceptual spaces. However, the approach to remote handling of fusion waste—which is necessary due to the fusion nuclear environment, as well as the quantities and the form of waste—may vary across reactor concepts. Accordingly, this subsection will detail the level and characterization of radioactive waste from fusion reactors, and then detail the challenges that come with managing that waste, focusing in particular on the development of remotely operated tools.

Waste from fusion reactors, much like waste from fission reactors, will require storage until the radioactive isotopes that have been produced in the operation of the reactor have decayed to levels whereby it is safe to dispose of the material. The main radioactive waste inventory results from materials activated by neutrons. Most of this is from the plasma-facing materials in the reactor. Some radioactive waste from a fusion reactor will be classified as low-level waste (LLW) within a 100-year timeframe, see [44]. However, depending on the materials chosen, plasma-facing materials such as divertor cassettes and first wall panels, both specific to MCF reactors, will remain radioactive for longer. These components will likely, in fact, be characterized as intermediate-level waste (ILW) which will require storage for a longer period [43, 44][13].

[12] Whilst I have previously argued that avoiding openness in discussing fusion waste is a PR-disaster-in-waiting for the fusion community—see [64]—this section, in keeping with the rest of the core sections of this chapter, will merely present information on fusion waste with the aim of informing the reader.

[13] In the UK, radioactive waste is characterized as LLW if both its alpha radiation activity is below 4 MBq kg^{-1} and if the sum of its beta and gamma activity is below 4 MBq kg^{-1} (or if the activity of beta or gamma, separately, is below 12 MBq kg^{-1}). ILW is anything above these limits that does not generate a significant amount of heat [43]. HLW is waste that may produce significant amounts of heat due to the radioactivity, which typically requires cooling. Fusion reactors are not expected to produce HLW, which typically comes only from fissile material such as spent nuclear (fission) fuel.

The half-life determines the length of time that waste is radioactive. ‘Long-lived’ waste arising from a fusion reactor is referred to as ILW due to the presence of isotopes with a long half-life. Such waste will not decay for hundreds or perhaps thousands of years. By important comparison, however, waste from a fission reactor may take hundreds of thousands or millions of years to decay. As noted, unlike fission reactors, fusion reactors can be designed to use materials that only result in short-lived radioactive waste. Certain materials produce longer-lived radioactive isotopes under irradiation, such as cobalt and chromium. Others can be designed to have low-activation characteristics, such as SiC and vanadium alloys [10, 44]. Judicious selection of materials and the development of effective low-activation materials can greatly reduce the concentration of long-lived isotopes that are produced. Low-activation materials either have low neutron absorption or have stable or radioactive products that are only activated for a short period, i.e. they have a short half-life. Even if materials are chosen judiciously, however, small levels of impurities can have a significant impact on the characterization of radioactive waste produced [43]. For instance, the uranium content in beryllium causes the production of plutonium, which is a long-lived and safeguarded radioactive isotope [45]. Special grades of beryllium have been developed with ultra-low uranium concentration to avoid this.

It is also important to place the characterization and quantity of fusion waste in context. A tokamak-type fusion reactor that uses RAFM steels and a lithium–lead breeder, for example, is shown to have a shutdown radiotoxicity (the level of radioactivity produced by the waste products immediately after shutdown) of about two orders of magnitude higher than a standard pressurized water fission reactor (PWR) [44]. This calculation includes the spent nuclear fuel content from a PWR, as well as components from, for example, the primary coolant circuit. The quantity of activated material—as in, the volume of the radioactive inventory to be managed—is also larger than a PWR [46]. So, in the immediate decommissioning phase of a fusion reactor, waste management activities may need to be much like those practised in the fission industry. What is different, and important, however, is that after a few hundred years—again, dependent on the fusion reactor and blanket materials selected, as well as whether the walls of the reactor will be self-healing—the waste from the fusion reactor will have decayed to a low enough level to be classified as LLW or as not radioactive at all (and thus disposable). Until that point is reached, however, fusion waste will require categorization and safe storage.

Another source of waste, which was mentioned in the last section of this chapter, is tritiated waste. Tritium produced in breeding blankets or from unburnt fuel will permeate through most in-vessel components. Treatment of irradiated and tritiated components requires hot cells to separate, recycle and dispose of tritium [43]. However, some tritium will remain in structural materials and breeding materials, even tritiated materials that are purged in the hot cells. Tritiated waste thus presents a technical challenge and design consideration, particularly for solid components such as tokamak divertors. Tritiated components, even if 95% of tritium is removed, will result in an increased inventory of ILW that must be treated as such for decades —this is true even if those components are not subject to neutron irradiation. This is because if the tritium is not (or cannot) be removed from a component, a period of 300 years is required for the radioactivity to decrease enough for it to be disposed of as non-radioactive waste [47].

On the face of it, then, the waste from a fusion reactor is a key cross-cutting issue across all fusion reactor concepts. While the development of flowing liquid blankets or first walls may negate full blanket replacement—which is a possible benefit of ICF and MIF reactors, in particular—even with thick liquid wall protection, the structure of the reactor will still become activated [48]. A divertor cassette is likely to be a highly irradiated component in an MCF reactor and the target debris from an ICF reactor will likewise produce several tonnes of irradiated material per year [48][14]. Flowing liquid surfaces in ICF and MIF reactors may mean that the walls of these types of reactor avoid becoming tritiated to the same levels as an MCF reactor, however. The extent to which tritium will permeate into the power generation systems—the heat exchangers and steam generators which transfer heat from the fusion reactor to create electricity or for process heat applications (as detailed subsequently)—is also an open question [37]. The tritium systems, in addition to servicing all components that become tritiated, will itself become tritiated. Therefore, the tritium systems must also be treated as waste at the end of life.

The nuclear environment in a fusion reactor—the radiation from neutrons, gamma rays, x-rays, and high-energy particles—prevents human maintenance and thus necessitates remote handling for the maintenance and replacement of components during its operation. Such remote handling equipment has to be capable of taking large payloads, gaining access to and maintaining movability in tight enclosures, and inspecting, removing (cutting), reattaching (welding), and cleaning components within the reactor [5, 49]. The equipment must have high availability for routine maintenance, and for cases of unexpected component failure and emergency repair [5].

Gamma radiation emitted by activated materials generates considerable dose rates during maintenance operations, meaning that all remote handling equipment will become irradiated and is likely to be damaged during maintenance operations. The high complexity of maintenance and the handling of irradiated components as well as irradiated remote handling equipment may form a large part of the operating cost of a fusion reactor [5]. The speed and performance of those remote handling systems to maintain equipment directly impacts the availability, and thus the development of such systems is a key factor in the economic success or failure of a commercial fusion reactor. Alongside other radiological protection measures akin to those on existing nuclear facilities, the remote handling systems are also critical for radiological protection.

As detailed previously in this section, while the characterization of waste, and the necessary management of that waste, is an issue that overlaps almost fully across concepts, the approach to designing remote handling systems varies. Remote handling systems for MCF reactors have, like many of the other challenges, received

[14] In principle, an MIF reactor using a liquid metal first wall (liner), with no fuel target and thus no debris, may present the concept with the lowest waste inventory.

by far the most attention and support from government-funded R&D. In particular, the development of remote handling technology for tokamaks have been conceptualized and some technologies tested. Remote handling experience on the JET tokamak has informed the design of the systems on ITER [50]. For a commercial fusion reactor, however, the systems for ITER—noting that it is an experimental reactor—will not be viable, due to the high cost, lengthy time taken to replace components, and reliability. This is because none of these factors were a key driver in the development of the technology for ITER. The complex geometries in a tokamak also mean that access to the inside of the reactor vessel is more complex. Detachable magnets that can be taken apart and put back together easily are being explored as a way of potentially avoiding having to work around such geometries [51]. Arguably, however, both the divertor and breeding blanket for MCF reactors present the greatest remote handling challenge. It is worth noting that this in part due to the fact that breeding blankets are expected to make up the majority of radioactive waste inventory of a reactor. The use of a lithium-lead breeding medium, for example, gives rise to ^{210}Po and ^{203}Hg under neutron irradiation—both of which are long-lived radioactive isotopes [52]. Large manipulators are also required to handle the heavy load of blankets. Both blanket and divertor systems must also be maintained after having been subjected to intense neutron irradiation. Neutron damaged steel is difficult to weld due to the voids created in the material microstructure [5]. Though these steel components could be replaced, the cutting and reconnecting of coolant pipe circuits, for example, is expected to present a challenge that may necessitate the removal and replacement of a greater number of parts at once [5]. Such technical challenges ratchet up complexity and serve to increase the cost and time taken to replace components. These factors might impact the commercial viability of a plant, including the required downtime for such complex maintenance. Quite obviously, for a commercial MCF reactor, blankets and divertors, in particular, must be developed with remote handling in mind. The diagram in figure 10.6 shows a proposed maintenance scheme for the blanket 'segments' and divertor 'cassettes' in a tokamak-type MCF reactor.

Although a greater number of conceptual designs and studies have been carried out for MCF reactors, as with blankets, ICF and MIF reactors may be advantageous as regards remote handling. The spherical chambers envisaged for most ICF and MIF reactors and the use of drivers or lasers, likely to be external to the main reactor vessel, may simplify maintenance. Despite this, some designs for ICF systems are not dramatically different from those envisaged for MCF reactors, as shown by the reactor vessel segment removal for the FI-DWRHL ICF concept in figure 10.7.

In some MCF concept designs, blanket systems are designed with the expectation that they must be replaced periodically when damage thresholds from heat and neutron loading are reached—a reasonable target for the lifetime of a blanket is five years or so. Solid-walled ICF and MIF reactors will likely face the same problems. However, as noted, the possibility of using liquid metals to act as a first wall and as a breeding medium could increase the lifetime of the reactor, and reduce the radioactive waste inventory, by preventing damage and, thus, early replacement [53]. Despite this, even with thick liquid wall protection, the structural materials of the

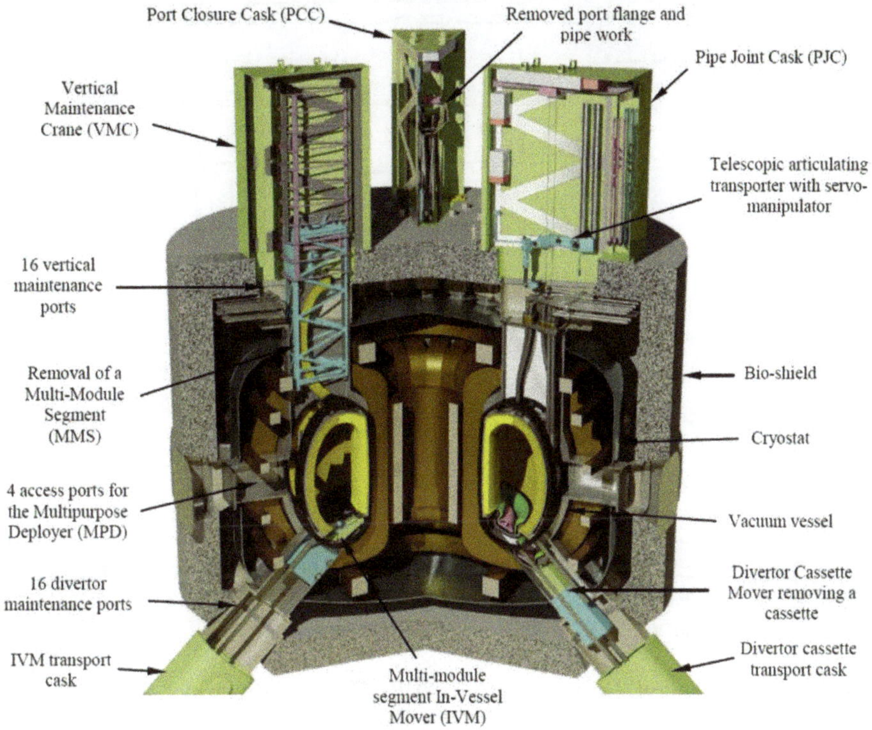

Figure 10.6. Remote handling systems to remove a blanket segment and divertor cassette on a tokamak-type MCF reactor. Reproduced from [49], copyright (2014), with permission from Elsevier.

reactor are still likely to become activated and damaged, albeit perhaps after a longer period of operation. Additionally, the production of unwanted isotopes in lead in any concept using a lithium–lead eutectic will still affect ICF and MIF reactors. One additional, possibly significant source of activation in ICF and some MIF reactors might occur from target debris (which, in some designs, could account for several tonnes of irradiated material per year) [48]. It is evident that ICF and MIF reactors will require systems for active maintenance and purification of any flowing liquid metal that is chosen for the design. As is the case for MCF reactors, waste is likely something that cannot be avoided—and is perhaps best viewed as a problem to be managed, rather than completely avoided.

Much less is known about the challenges ahead for designing remote handling systems for MIF reactors, and fewer designs have been developed. Despite this, to some extent, designs can lean on approaches developed for MCF and ICF reactors. Additional complexity may arise from, for example, the use of diagnostics inside the reactor vessel compared with MCF reactors. As detailed in chapter 2, the environment in an ICF reactor for diagnostics places a different constraint on developing, and subsequently on maintaining and replacing, suitable technologies compared with MCF reactors. The same is true for MIF reactors. Therefore, more detailed technical assessments are needed to fully understand the challenges.

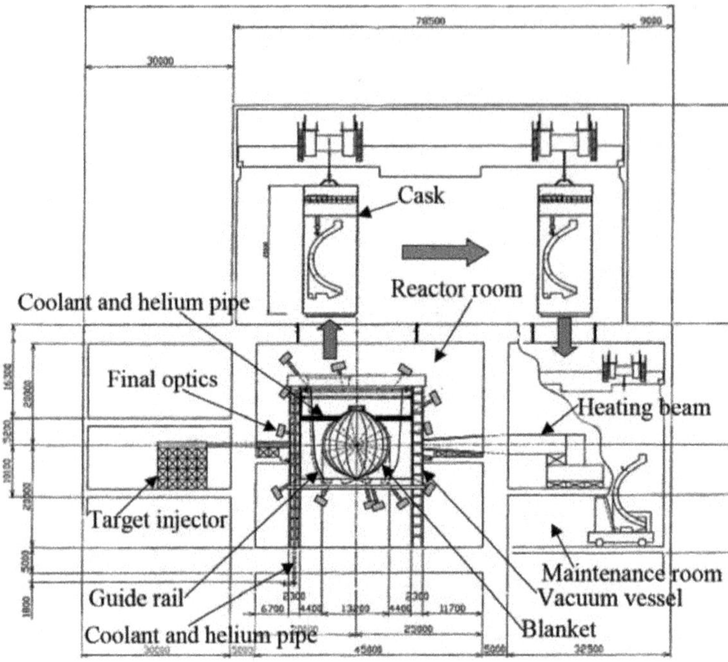

Figure 10.7. Remote maintenance layout for the FI-DWHRL conceptual ICF reactor. Reproduced from [19] with permission of the American Nuclear Society (http://www.ans.org/).

10.2.5 Balance of plant systems

Balance of Plant (BoP) refers to all the systems and buildings required for the operation of the fusion reactor. BoP systems are often separated into several categories. For simplicity, in this chapter, the BoP systems are considered as something of a coverall term to describe all power conversion systems; all systems to supply electrical power; and all buildings, structures, or technologies required for the successful operation of a fusion power plant. Many of the components or even whole systems that have been mentioned in this chapter (and in chapter 2) interface with BoP systems in some way. The lasers and drivers require electrical power, for instance, and must be connected to power supplies. The blanket heat exchanger must be connected to turbine systems to generate energy from the heat, which is supplied by the neutrons created by the fusion reaction. The tritium systems (including the hot cells) must be connected to ventilation and radiation monitoring systems for plant safety. The schematic in figure 10.8 shows these systems for a conceptual tokamak power plant (DEMO).

First, let us consider the power conversion systems. A key system in any fusion reactor power plant is the primary heat transfer system, which converts the thermal power (supplied by the blanket) into usable energy via a form of heat exchanger. The BoP power conversion systems thus include the heat exchanger (connected to both the blanket and the power exhaust), high-pressure and low-pressure turbines, cooling towers, and associated circuitry to connect these systems. The major difference between MCF and ICF or MIF regarding the primary heat transfer system is that the

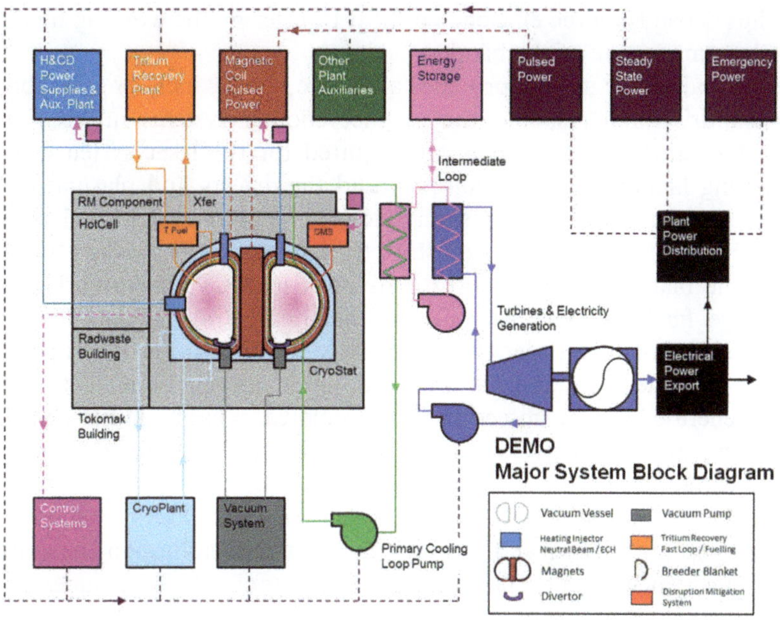

Figure 10.8. Tokamak plant layout for a DEMO tokamak-type reactor. Reproduced from [54] under a CC-BY license.

former relies on a steady-state operation, whereas the latter two rely on pulsed operation. However, such pulsed machines will be capable of producing a continuous power output by design if the pulses are frequent enough [55]. In fact, some MCF designs are not dissimilar to ICF and MIF designs, as they account for the dwell periods between plasma pulses (albeit that such pulses are far longer than for ICF and MIF concepts), to allow for continuous and uninterrupted power generation—which is a desirable feature of any energy technology [56]. Accordingly, whether steady-state or pulsed, fusion reactor power can be transferred continuously to the power conversion systems or can be accumulated in an energy transfer and storage (ETS) system. More important than whether reactors are steady-state or pulsed is that the majority of fusion power plant concepts are ostensibly focused on electricity generation (also see chapter 7). Typically, concepts are water-cooled and envisage a standard steam cycle (Rankine cycle), but helium, as well as self-cooled liquid metal or molten salt concepts, might facilitate thermal power cycles—including a Brayton cycle—to achieve higher efficiency [56, 57][15,16]. As such, BoP systems for an electricity-generating fusion power plant are typically considered to be akin to other steam power generation systems for electricity production. The efficiency of such

[15] Due to the pumping power required for helium-cooled systems, the expected electrical power required is about double that of water-cooled systems—but this difference might be negated by the higher efficiency afforded by operating at the higher temperatures afforded by helium.

[16] Helium for use in future fusion reactors has been the subject of several studies in recent years. Such studies place particular focus on the volatility of the supply of helium as a natural resource. See [58, 59, 60].

systems thus depends on the efficiency of heat transfer by the working fluid and—of course—the temperature of the blanket.

The electrical power system provides all onsite power, including for the heating and current drive systems, and the magnet cooling systems in MCF reactors. Similarly, the same power systems are required for the laser systems and target manufacturing facility in ICF reactors[17], and the drivers and plasma injectors in MIF reactors. Across all concepts, the electrical power system will provide the power for the control systems, diagnostics, tritium systems, as well as for the active cooling of the blanket systems (and the divertor in MCF reactors), hot cells and waste storage facilities. The recirculating electrical power in an MCF plant is estimated to be about one order of magnitude larger than for a fission power plant. This means that the site might require a well-interconnected electricity grid unless electricity generated can be immediately recirculated, reliably. ITER has a reactive power compensation system to ensure that the power absorbed from the grid does not exceed certain levels.

Finally, the auxiliary systems are here taken to include cold water supply, which is required for reactor cooling and power generation (depending on the concept), as well as for emergencies—in the case of a fire, for example. It also includes storage for compressed gas, fuel and spare components. It also covers all onsite infrastructure and buildings, including the control room, administrative buildings, security and access, laboratories and workshops, hot cells, and the waste storage building [55, 56]. An alternative layout of BoP systems for a generic fusion power plant, showing many of these auxiliary systems, is provided in figure 10.9.

As the current focus in fusion development is—understandably—focused on the development of the fusion power core, i.e. the fusion reactor itself, the design and development of BoP systems for a fusion power plant remains at the conceptual stage. ITER provides many of the auxiliary BoP systems required to operate a fusion reactor, but ITER is not a power-producing fusion reactor and is missing several key systems that would be present in a commercial plant. Several conceptual designs for the layout of a fusion power plant have been developed. They have previously been based on fission power plant layouts. BoP systems in fission power plants can account for a substantial portion of the cost of the overall plant. The LIFE reactor costs are predicated on equivalent fission reactor costs, whereby the fission reactor core is replaced by the ICF reactor core in the cost estimates to give a crude a cost estimate of between US$2 B and US$6 B (2008 US$) [57]. Other BoP systems for MCF fusion plant concepts are more comprehensive (see [55, 56]). Fortunately, many of the BoP systems required for a fusion reactor system based on the D–T fusion cycle are broadly the same, and thus the differences across concepts are negligible. Despite this, there are likely to be some distinct differences between some concepts. Altogether, BoP systems are generally at a high level of technology readiness, but require adaption for a fusion power plant. The challenge is to design a fusion power plant that is simultaneously high performance, safe and cost-effective.

[17] The power required for a laser is a significant, so reducing the amount of energy to power ICF reactors is a key focus of ICF development programmes. See chapter 2.

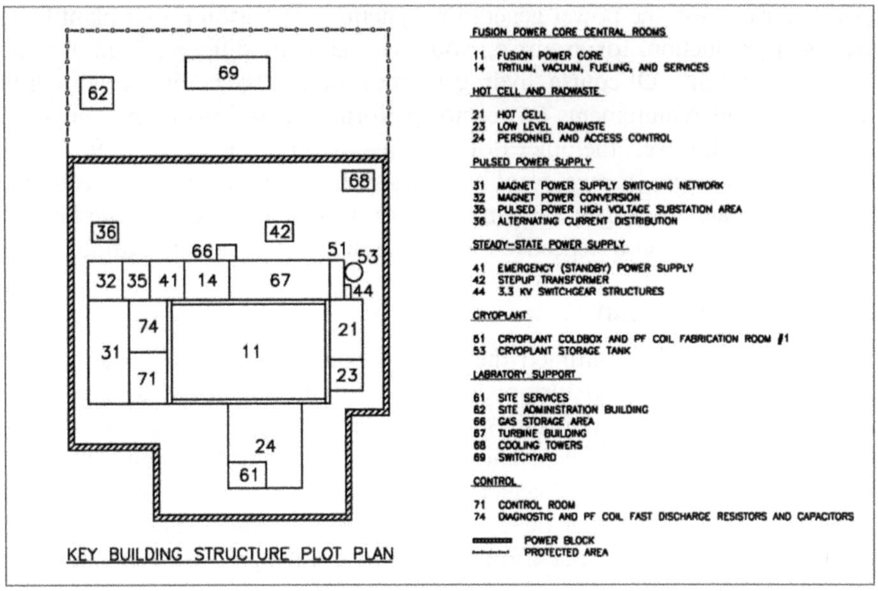

Figure 10.9. Conceptual layout for a fusion electricity generation power plant. Image provided courtesy Simon Woodruff, from work conducted under grant from ARPA-E DE-AR0001175 'Cost analysis of four compact modular fusion power cores' PI: Simon Woodruff (2020).

Finally, it is instructive to again consider the cost of BoP systems for an equivalent fission power plant as a benchmark. In addition to the operating costs and the additional costs required for safety and security, the cost of the fission reactor itself (not including the rest of the plant equipment) sets the bar for the cost of the fusion reactor core. If crudely assuming that all BoP systems are equal cost for a fusion and fission power plant, then all other systems and components that make up the fusion reactor, i.e. all systems that have been detailed previously in this chapter (the first wall, blanket, remote handling equipment etc), as well as those in chapter 2, must be similar to the cost of a fission reactor core to be commercially viable. Compared with a PWR core, which is relatively simple, a fusion reactor core is thus incredibly complex. Such high complexity could render fusion substantially more expensive than fission. The fundamental point in contention here is that fusion must compete economically with other modes of electricity generation. Unless the reactor core can be simplified tremendously—as is indeed the goal envisaged by several private fusion companies, including those focused on the tokamak, as detailed in chapter 7—based on this somewhat oversimplified but useful indicative cost comparison, it is hard to understand how an ITER-scale reactor would ever be economical. For reference, although the machine was designed for experiments and, inter alia, the machine incorporates a wealth of expensive scientific equipment for experimentation, such as diagnostics, ITER does not employ a full breeding blanket and has only a few of the required BoP systems required for a commercial reactor, yet the current projected cost is around US$ 25 B. Even if a next-step reactor were to cost even half that amount, it is uncertain whether it would be cost-competitive.

Finally, for interest, the power generation systems for a fusion power plant focusing on hydrogen production, for example, would be markedly different from a plant for electricity generation[18]. Of course, hydrogen production as the commercial application will place different requirements on reactor performance and operating temperatures (see chapter 11). However, the other BoP systems—as shown in figure 10.9, e.g. power supplies, compressed gas storage, plant buildings such as the control room, security, and access etc—are likely to largely remain the same. Hence, BoP systems are largely the same across concepts and irrespective of the chosen commercial application.

10.3 Fusion innovation

The previous section has outlined the broad engineering challenges that must be solved to realise commercial fusion. None of the challenges can be avoided, and some are easier to resolve—depending on the fusion reactor concept. All challenges have varying degrees of technical uncertainty and will require significant cost and time to develop. To effectively accelerate progress will require a strategy that does not consider fusion technology development inside a vacuum. The final section of this chapter thus explores the aforementioned challenges from the innovation perspective, by explicitly considering commercial drivers.

10.3.1 Seeking successful fusion innovation

Fusion R&D programmes necessarily foresee the ramping up of parallel R&D programmes to solve the technical challenges that have been briefly outlined in this chapter. However, such programmes, the vast majority of which have historically been led by government-funded laboratories, has been to an extent motivated by advancing human understanding of fusion science. As detailed at the start of this chapter, the approach to fusion innovation has been for technological innovation—the demonstration of technology for scientific gain—rather than commercialisation. The scaling-up and market penetration of a new technology—in particular, a new energy technology—can take decades for a variety of technological, infrastructural, sociological, and financial reasons [3, 61]. The conditions of the external environment in which the technology sits must be considered early in development to ensure the swiftest route to market[19]. Here, I suggest an approach to understand this environment based on the identification and characterisation of commercial drivers. Commercial drivers are here defined as anything that might influence the development of a technology being turned into a product destined for commercial market[20].

[18] It is also worth mentioning that if the fusion industry were to pursue such alternative, advanced markets—which may be unobtainable or undesirable for competitor technologies such as wind and solar—it could unlock a different path to successful commercialisation; one that is not based on its adherence to the varying stringent requirements of the electricity generation market.

[19] New technology can also drive the market behaviour and requirements, if it offers something novel and/or desirable, but the important point here is that the two are—and should be considered as—coupled.

[20] As shall become clear subsequently, when I use the term 'commercial', here, I do not necessarily mean for this to be taken as synonymous with 'for financial gain'. In the modern world, the viability of a technology depends on much more than just cost, and a product that is more expensive but which provides general societal benefit, can still succeed.

To some extent, commercial drivers have been inherent in fusion development. Systems for fusion reactor concepts are being designed with reliability (see materials), maintenance (see remote handling) and high-efficiency energy generation (see blankets) in mind. For example, the LIFE project—an ICF reactor that was recently abandoned, principally due to technical difficulties (it was derived from NIF, which is no longer regarded as a likely route to commercial fusion energy)—is just one of several initiatives that considered commercial drivers in its design. LIFE conceptual studies considered the capital investment need, engagement with authorities responsible for licensing and with electricity utilities, and even 'timeliness of delivery' as commercial factors, tightly coupling them to the technology development [62][21]. To be successful in commercialisation, we must follow on from this—and expand it—to understand the space in which technology is both technically and commercially viable. I refer to this as the *fusion innovation space*. Private fusion developers, who are already inherently focused on commercialisation—see [1]—have an opportunity to identify their own fusion innovation space. Doing so will allow them to identify and then pursue technology development with both the technical and commercial drivers in mind when they tackle the challenges outlined in this chapter.

10.3.2 Commercial drivers for fusion

Commercial drivers can impact fusion development efforts in a variety of ways. Obvious commercial drivers include component or system lifetime, plant capacity factor (or product output, in the case of a fusion reactor for non-electrical applications), the raw material and manufacturing costs, or the cost of electricity (to determine competitiveness). Less tangible commercial drivers, however, are also important. Such drivers might include the cost of financing; the supply of labour; and laws and regulations for reactor siting, radiological protection, or even things that may seem trivial by comparison such as public perception, or levels of staff satisfaction and mental wellbeing. Again, such considerations are likely already considered as drivers under the umbrella of corporate governance at large fusion programmes, including at government-led laboratories. However, these drivers must be connected to technology development. An effective tool to identify drivers, and to then connect them to technology development, is PESTLE analysis.

PESTLE analysis is a simple but expansive tool that can be used to identify external commercial drivers—political, economic, social, technological[22], legal, and environmental; hence 'PESTLE'—that affect a particular organisation, mission, or even a technology or component. As such, PESTLE analysis can be effectively deployed at various levels. It can be applied to a specific issue, for example, the development of SiC/SiC material for advanced breeding blankets; at a system level,

[21] Other examples are the FESAC (Fusion Energy and Sciences Advisory Committee) report [68] and the European tokamak 'DEMO' programme [69].

[22] The 'T' (for 'technological') in PESTLE refers specifically to external technological drivers that might impact an internal development programme. For instance, this might be the technical challenge of qualifying a new nuclear material that may have come from the aerospace sector.

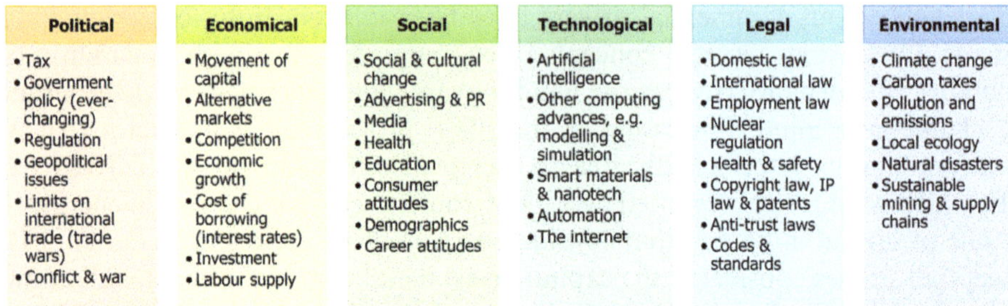

Figure 10.10. Examples of PESTLE drivers for the development of fusion technology, based on previous research by the author in [63].

for example, for the development of the driver for an MIF liner; or even to the overall mission level, for example, to understand the effects of pursuing hydrogen production, rather than electricity generation. The diagram in figure 10.10 shows a broad range of commercial drivers for the development of commercial fusion technology.

PESTLE analysis, integrated as part of an innovation strategy, can help to align technology development with commercial needs and can help identify a viable fusion innovation space. To underpin the utility of the technique, three examples of challenges that were mentioned in the earlier sections of this chapter are briefly explored via a PESTLE analysis. The following issues are all key challenges to be overcome to achieve commercial fusion, and are all linked.

Several studies attempt to estimate a fusion cost of electricity (CoE), but these are projected 20 to 30 years into the future when what is considered 'competitive' is difficult to forecast. Even if the first commercial fusion reactor for electricity is delivered within the next decade, it must compete with well-established energy sources and other low-carbon sources that will continue to improve[23]. While defining allowable CapEx and OpEx provides a crude but useful target for an electricity-producing fusion power plant, it is instructive to consider the potential of non-electricity markets for fusion technology development instead. Alternative markets may allow fusion to fill a niche in which it plays to the technology's strengths and—importantly—does not have to, for example, compete with advanced solar, wind, nuclear, or even fossil fuels with carbon capture and storage. What would the application of fusion for hydrogen production or desalination mean for the required performance of the technology? Firstly, a fusion power plant that is required to produce hydrogen or to desalinate water does not have to operate at the same capacity factor (how often the plant is 'up and running'). Further, the fusion plant is producing a product that can be stockpiled and sold. So, the viability of the fusion power plant is not in its ability to operate continuously (something that is achieved by nuclear fission reactors and closed-cycle gas turbines and very much valued by grid operators). The lifetime of materials in a plant for desalination that is only required to run at, say, 50% capacity, means that components could last roughly twice as long (challenges associated with thermal cycling notwithstanding).

[23] Also see [61].

Moreover, some BoP systems such as steam turbines are not necessary to produce the product, which would facilitate a saving in the efficiency of thermal to electricity conversion. From a legal perspective, a fusion plant for desalination does not have to provide guarantees to grid operators that it will provide continuous power. Instead, it will have to adhere to a separate set of laws that govern clean water production (in the case of a fusion desalination plant), for example. A political factor is that grants may be available from governments looking to find solutions to decarbonise non-electricity sectors—the production of clean hydrogen may appear attractive in this regard. All such factors provide a commercial fusion power plant for these applications with much greater flexibility, and the option for such commercial pathways should be steadfastly explored, perhaps using this PESTLE analysis as a starting point.

Next, consider the challenge associated with the use of next-generation fusion materials. As alluded to earlier in this chapter, the development of structural materials for fusion—and indeed the development of materials for any energy technology—is a long and costly process. In part, this is because of the period required to develop a new material followed by an even longer testing period, which is first to validate its performance in an application-relevant environment, and then to qualify the material with the appropriate licensing authorities [7]. Materials are required to achieve high performance, but at what cost is this acceptable? Choosing expensive materials on the premise of high performance alone is not conducive to a commercially viable power plant (unless such materials can be developed at a truly low cost). Advanced steels may prevent the production of higher levels of long-lived radioactive waste. However, let us consider the social aspects of choosing another readily available material, but which yields long-lived radioactive waste. Let us assume this existing material is already qualified for nuclear application, i.e. it is licensed, and ready to be used in our fictitious fusion reactor without regulatory issue. Naturally, the use of such a material in the fusion environment will require specialized handling equipment to managed safely at the end of life. However, from an environmental and economic perspective, the existing material—let us imagine it is a grade of steel—may have an established international supply chain that can deliver the material at low cost and on rapid timescales. Compare this to another newly developed steel that may use a more expensive isotope for alloying, which is mined in smaller quantities and does not yet have an existing supply chain. Taking a holistic approach, and viewing this choice through the lens of PESTLE analysis, we find that perhaps the most pertinent issue in employing the existing steel over the novel steel is the impact on public acceptance. Though it is important to note that even a fusion reactor with the most judiciously selected materials will produce radioactive waste, here there is a trade-off of technical performance, economics, and some environmental and legal factors, against social or perhaps political factors. Indeed, the concept of obtaining a 'social license' (see [64]) is important more broadly for the development of fusion. However, it could be argued that obtaining social acceptance for the use of a material that will result in long-lived radioactive waste, or, as another example, highly tritiated waste—both of which require specialized handling and waste management procedures—would be acceptable to the public if the fusion reactor were to offer a beneficial solution to many other social problems (clean air, clean water etc). Again, viewing the materials issue through such lenses—not just the technical, or the waste (environmental)—provides a perspective that enables a technologically and commercially viable fusion innovation space to be identified.

As a final example, let us briefly consider the supply of beryllium resource for a tritium breeding blanket. Beryllium is scarce on Earth, and the supply chain for the material is significantly limited—one large fusion reactor requires an amount that is approximately the same as the total global quantity of beryllium mined and machined per year—see [63, 65]. Viewing the problem through different lenses, we find that there are commercial challenges. Through an economic lens, we see that beryllium is expensive. Similarly, through an environmental and a political lens, the scaling-up of the beryllium industry would require the expansion of mining operations and may require government guarantees to the beryllium industry [63]. When we consider an alternative solution to the problem—here, lead-based blankets—we find similar environmental and social challenges, since, as detailed previously, lead produces long-lived isotopes of mercury and polonium which necessitates a dedicated waste management strategy [52]. In both cases, once again, we find ourselves in a position where we have a trade-off that requires a strategic decision based on commercial, not technological, drivers.

10.3.3 An innovation strategy for cooperative public and private sector fusion development

Of course, the intention here was not to fully dive into the question of whether hydrogen is the best market application for fusion or whether the fictional steel in question was better than an also fictional—but presumed to exist—nuclear-grade steel. Instead, it is to illustrate the utility of applying PESTLE analysis to provide a clearer view of the pathway to commercial fusion and to show the benefit of considering how to develop an innovation strategy to get there[24]. Commercial drivers must be considered in solving the technical challenges that have been detailed throughout this chapter.

Commercial challenges that affect the overall mission, such as licensing and regulation, or indeed materials development are—to a large extent—cross-cutting issues. As such, there is a need to further delineate between the challenges that must be solved by private fusion developers, and by the broader fusion community (in particular, government-funded laboratories). As noted, private fusion developers are better suited to accelerating the pace of fusion commercialisation, principally due to an agile innovation approach [1]. However, here we can tie this philosophy to the technical challenges that have been detailed in this chapter to give the broad outline of a public-private innovation strategy.

[24] A framework for the development of such an innovation strategy with the support of Technology Roadmapping can be derived from the author's previous publication [1].

Private fusion companies are hoping to press forward in making rapid progress towards a working fusion reactor, perhaps within the next decade (see chapter 7). Even if such a reactor is technically and commercially imperfect, it can be improved over time. However, these companies do not have the budget, capability, or the time to develop novel materials, develop specific BoP systems, or even develop tritium breeding blankets. Solutions to next-step challenges must be developed to be commercially viable. Public laboratories, as well as fusion engineering companies, should aim to match the pace and ambition of private fusion developers and should focus specifically on developing and providing solutions to well-known problems, many of which have been detailed in this chapter. In doing so, they should explicitly consider commercialisation, so that they can support private industry in its development efforts whilst simultaneously continuing research in support of existing publicly-supported scientific experiments such as ITER. Strategic partnerships across organizational boundaries will result in the creation of a new fusion ecosystem that can spark a new wave of fusion innovation, but this time angled towards commercialisation. In such an ecosystem, public laboratories, as well as universities and research institutions, can move into a role in which they provide world-class scientific R&D to support private industry, instead of designing and developing the programmes themselves. Fusion laboratories can angle themselves to partner with private organizations to support commercial development by carrying out world-class R&D, in much the same way that nuclear laboratories support advanced nuclear fission development.

None of the above should preclude public science programmes, including ITER, from being continually supported to advance scientific understanding. However, such projects should not be pursued on the premise that they will lead to commercial fusion energy. Instead, laboratories should endeavour to become akin to NASA, whilst private fusion companies should strive to become like SpaceX [1]. Private companies can, therefore, focus on developing their technologies for commercial exploitation, adopting an approach to innovation that is not possible in government-funded programmes. Entrepreneurs can do what they do best: drive commercial exploitation, and the world's leading scientists can support them with advanced technological innovation!

10.4 Conclusions

The contribution of this chapter is twofold: firstly, it provides a high-level overview of the technology challenges facing almost all fusion development efforts, and secondly, it explicitly considers the drivers that must be considered for successful fusion innovation, and ultimately for fusion commercialisation. More specifically, it paints a picture of where we want to be and describes the steps—identifying the key challenges and potential solutions to those challenges—to get there. In understanding the broader technical challenges, it is shown that to realise commercial fusion requires a wholly interdisciplinary approach. Moreover, it shows there is some scope for cross-cutting development in solving next-step challenges; more than in the development of technology for specific reactor configurations (as outlined in chapter 2). However, it has also shown that various players across the fusion ecosystem must tackle all challenges—indeed, it highlights explicitly that all actors in that system must be focused more sharply on commercialisation, not just on developing technology for technology's sake.

Of particular importance is the challenge presented by the development of fusion materials. Materials underpin all the challenges detailed in this chapter in one way or another. There is an acute need to develop viable tritium breeding blankets, as a system that is directly related to the commercial viability of a fusion reactor with the dual function of breeding fuel and producing power. Neither materials for fusion reactors nor fusion blankets have been demonstrated. Greater focus is needed in these areas to deliver fusion in the near-term (within decades). Although the plasma-facing components and the tritium breeding blanket are the most obvious examples of materials challenges, many of the challenges associated with diagnostics, for instance, are related to the lifetime of the components, which is fundamentally a materials issue. Similarly, waste management considerations and remote handling strongly impact the lifetime of components. If it were possible to dream up materials that achieved the perfect performance for each of these, then many of the challenges would largely be solved.

Finally, all fusion efforts focused on realising commercial fusion, regardless of the specific reactor design, must traverse the engineering challenges outlined in this chapter. All efforts must start with testing on current experimental reactors, before advancing them further until the technologies are ready to be integrated in a first commercial prototype. None of the challenges detailed here is insurmountable. A new approach is required; one in which innovation is conducted through building and testing on rapid cycles, using the best possible technologies, rather than aiming for the optimal solution. A first-pass technology can later be improved upon, but we must focus on actually building and testing that technology—instead of trying to develop a perfect solution from day one. Taking a different perspective on the analogy of the Wright Brothers used in chapter 7—it is again noteworthy that their first 1903 flyer did not look like a modern jet aircraft; nor did the Windscale pile or the EBR-I fission reactors look anything like a modern-day PWR. A first energy demonstration fusion reactor should not strive to be perfect. Instead, it should show proof of principle, and—importantly—must scale to a commercial reactor. This chapter provides a view of some of the key challenges and technologies that must be developed to overcome those challenges. A commercial fusion reactor will only be realised if all of these can be solved, and all solutions integrated. A dedicated approach is needed to solve all the problems concurrently and for commercial exploitation. Many of the solutions to the challenges ahead are concept-specific, but several are cross-cutting. The fusion ecosystem needs to be upended to allow the new world of private fusion companies to flourish and to drive innovation in both the public and private sector, who should work more closely, to accelerate the realisation of commercial fusion. We must make a new promise to make fusion a reality in the next 30 years. To fulfil this promise, which has not been kept before now, fusion must enter a new age of innovation.

References

[1] Pearson R J, Costley A E, Phaal R and Nuttall W J 2020 Technology Roadmapping for mission-led agile hardware development: a case study of a commercial fusion energy start-up *Technol. Forecast. Soc. Change* **158** 120064

[2] O'Bryan M 2013 Innovation: the most important and overused word in America *Wired* https://www.wired.com/insights/2013/11/innovation-the-most-important-and-overused-word-in-america/ (accessed 12 November 2020)

[3] Gallagher K S, Holdren J P and Sagar A D 2006 Energy-technology innovation *Annu. Rev. Environ. Resour.* **31** 193–237

[4] Stork D and Zinkle S J 2017 Introduction to the special issue on the technical status of materials for a fusion reactor *Nucl. Fusion* **57** 92001

[5] Buckingham R and Loving A 2016 Remote-handling challenges in fusion research and beyond *Nat. Phys.* **12** 391–3

[6] Zinkle S J and Ghoniem N M 2000 Operating temperature windows for fusion reactor structural materials *Fusion Eng. Des.* **51** 55–71

[7] Zinkle S J and Busby J T 2009 Structural materials for fission & fusion energy *Mater. Today* **12** 12–9

[8] Rubel M 2019 Fusion neutrons: tritium breeding and impact on wall materials and components of diagnostic systems *J. Fusion Energy* **38** 315–29

[9] Tan L, Snead L L and Katoh Y 2016 Development of new generation reduced activation ferritic-martensitic steels for advanced fusion reactors *J. Nucl. Mater.* **478** 42–9

[10] Perlado J M and Sanz J 1993 Neutron damage and activation of the first wall of inertial confinement fusion reactors: recycling and waste disposal *Laser Part. Beams* **11** 437–42

[11] McCracken G and Stott P 2013 *Fusion: The Energy of the Universe.* (Waltham, MA: Academic)

[12] Giancarli L *et al* 2000 R&D issues for SiCf/SiC composites structural material in fusion power reactor blankets *Fusion Eng. Des.* **48** 509–20

[13] Snead L L, Nozawa T, Ferraris M, Katoh Y, Shinavski R and Sawan M 2011 Silicon carbide composites as fusion power reactor structural materials *J. Nucl. Mater.* **417** 330–9

[14] Nozawa T *et al* 2018 Japanese activities of the R&D on silicon carbide composites in the broader approach period and beyond *J. Nucl. Mater.* **511** 582–90

[15] El-Guebaly L, Mynsberge L, Menard J, Brown T, Malang S and Waganer L 2014 Nuclear aspects and blanket testing/development strategy for ST-FNSF *IEEE Trans. Plasma Sci.* **42** 1457–63

[16] Knaster J, Moeslang A and Muroga T 2016 Materials research for fusion *Nat. Phys.* **12** 424–34

[17] Odette R and Zinkle S 2019 *Structural Alloys for Nuclear Energy Applications.* (Newnes)

[18] IDO S *et al* 1982 Neutronics calculations in pellets and blankets of laser fusion reactor concept SENRI-I *J. Nucl. Sci. Technol.* **19** 1019–36

[19] Hiwatari R *et al* 2007 Preliminary consideration on maintenance approach for a fast ignition ICF reactor with a dry wall chamber and a high repetition laser *Fusion Sci. Technol.* **52** 911–5

[20] Humrickhouse P W, Merrill B J, Yoon S-J and Cadwallader L C 2019 The impacts of liquid metal plasma-facing components on fusion reactor safety and tritium management *Fusion Sci. Technol.* **75** 973–1001

[21] Nakai S and Mima K 2004 Laser driven inertial fusion energy: present and prospective *Reports Prog. Phys.* **67** 321
[22] Miller R L 2007 Perspectives on magnetized target fusion power plants *J. Fusion Energy* **26** 119–21
[23] Laberge M *et al* 2013 Acoustically driven magnetized target fusion *2013 IEEE 25th Symp. on Fusion Engineering (SOFE)* pp 1–7
[24] Guo H *et al* 2020 Innovative approaches towards an economic fusion reactor *Natl Sci. Rev.* **7** 245–47
[25] Pearson R J, Antoniazzi A B and Nuttall W J 2018 Tritium supply and use: a key issue for the development of nuclear fusion energy *Fusion Eng. Des.* **136** 1140–8
[26] Kovari M, Coleman M, Cristescu I and Smith R 2018 Tritium resources available for fusion reactors *Nucl. Fusion* **58** 026010
[27] Ricapito I, Cismondi F, Federici G, Poitevin Y and Zmitko M 2020 European TBM programme: first elements of RoX and technical performance assessment for DEMO breeding blankets *Fusion Eng. Des.* **156** 111584
[28] Abdou M *et al* 2015 Blanket/first wall challenges and required R&D on the pathway to DEMO *Fusion Eng. Des.* **100** 2–43
[29] Fischer U *et al* 2016 Neutronic performance issues of the breeding blanket options for the European DEMO fusion power plant *Fusion Eng. Des.* **109–111** 1458–63
[30] Tillack M S, Humrickhouse P W, Malang S and Rowcliffe A F 2015 The use of water in a fusion power core *Fusion Eng. Des.* **91** 52–9
[31] Enoeda M *et al* 2003 Design and technology development of solid breeder blanket cooled by supercritical water in Japan *Nucl. Fusion* **43** 1837
[32] Ihli T *et al* 2008 Review of blanket designs for advanced fusion reactors *Fusion Eng. Des.* **83** 912–9
[33] Forsberg C, Zheng G, Ballinger R G and Lam S T 2019 Fusion blankets and fluoride-salt-cooled high-temperature reactors with flibe salt coolant: common challenges, tritium control, and opportunities for synergistic development strategies between fission, fusion, and Solar Salt Technologies *Nucl. Technol.* 1–24
[34] Hernández F A *et al* 2019 An enhanced, near-term HCPB design as driver blanket for the EU DEMO *Fusion Eng. Des.* **146** 1186–91
[35] Okino F, Frances L, Demange D, Kasada R and Konishi S 2017 Tritium recovery efficiency from an array of PbLi droplets in vacuum *Fusion Sci. Technol.* **71** 575–83
[36] Malang S, Tillack M, Wong C P C, Morley N and Smolentsev S 2011 Development of the lead lithium (DCLL) blanket concept *Fusion Sci. Technol.* **60** 249–56
[37] Konishi S, Glugla M and Hayashi T 2008 Fuel cycle design for ITER and its extrapolation to DEMO *Fusion Eng. Des.* **83** 954–8
[38] Reyes S *et al* 2013 Life tritium processing: a sustainable solution for closing the fusion fuel cycle *Fusion Sci. Technol.* **64** 187–93
[39] Laberge M 2009 Experimental results for an acoustic driver for MTF *J. Fusion Energy* **28** 179–82
[40] Sawan M E and Abdou M A 2006 Physics and technology conditions for attaining tritium self-sufficiency for the DT fuel cycle *Fusion Eng. Des.* **81** 1131–44
[41] Federici G *et al* 1999 In-vessel tritium retention and removal in ITER *J. Nucl. Mater.* **266** 14–29

[42] Boyer C *et al* 2009 Tritium in plants: a review of current knowledge *Environ. Exp. Bot.* **67** 34–51

[43] Gilbert M R, Eade T, Bachmann C, Fischer U and Taylor N P 2017 Activation, decay heat, and waste classification studies of the European DEMO concept *Nucl. Fusion* **57** 46015

[44] Zucchetti M, Candido L, Khripunov V, Kolbasov B and Testoni R 2018 Fusion power plants, fission and conventional power plants. Radioactivity, radiotoxicity, radioactive waste *Fusion Eng. Des.* **136** 1529–33

[45] Kolbasov B N, Khripunov V I and Biryukov A Y 2016 On use of beryllium in fusion reactors: resources, impurities and necessity of detritiation after irradiation *Fusion Eng. Des.* **109–111** 480–4

[46] El-Guebaly L, Massaut V, Tobita K and Cadwallader L 2008 Goals, challenges, and successes of managing fusion activated materials *Fusion Eng. Des.* **83** 928–35

[47] Gerchikov M, Glodeanu F, Patrascoiu S and Naum M 2015 Why a TRF shall be built on cernavoda site *Progress Cryog. Isot. Sep.* **18** 17–26

[48] Conn R W, Najmabadi F and Tillack M S 1998 Prospects and issues for commercial fusion power systems *Fusion Eng. Des.* **41** 337–47

[49] Loving A *et al* 2014 Pre-conceptual design assessment of DEMO remote maintenance *Fusion Eng. Des.* **89** 2246–50

[50] David O, Loving A B, Palmer J D, Ciattaglia S and Friconneau J P 2005 Operational experience feedback in JET remote handling *Fusion Eng. Des.* **75–79** 519–23

[51] Sorbom B N *et al* 2015 ARC: a compact, high-field, fusion nuclear science facility and demonstration power plant with demountable magnets *Fusion Eng. Des.* **100** 378–405

[52] Taylor N *et al* 2017 Materials-related issues in the safety and licensing of nuclear fusion facilities *Nucl. Fusion* **57** 92003

[53] Electric Power Research Institute 2012 *Program on Technology Innovation: Assessment of Fusion Energy Options for Commercial Electricity Production* (Palo Alto, CA: Electric Power Research Institute) Report no. 1025636

[54] Federici G, Biel W, Gilbert M R, Kemp R, Taylor N and Wenninger R 2017 European DEMO design strategy and consequences for materials *Nucl. Fusion* **57** 92002

[55] Bechtel National Inc., Woodruff Scientific Inc., and Decysive Systems 2017 Conceptual cost study for a fusion power plant based on four technologies from the DOE ARPA-E ALPHA Program (https://researchgate.net/deref/http%3A%2F%2Fdx.doi.org%2F10.13140%2FRG.2.2.24116.55688)

[56] Ciattaglia S, Federici G, Barucca L, Lampasi A, Minucci S and Moscato I 2017 The European DEMO fusion reactor: design status and challenges from balance of plant point of view *2017 IEEE Int. Conf. on Environment and Electrical Engineering and 2017 IEEE Industrial and Commercial Power Systems Europe (EEEIC/I&CPS Europe)* pp 1–6

[57] Meier W R *et al* 2009 Systems modeling for the laser fusion-fission energy (LIFE) power plant *Fusion Sci. Technol.* **56** 647–51

[58] Nuttall W J, Clarke R and Glowacki B 2012 *The Future of Helium as a Natural Resource* ed W J Nuttall, B A Glowacki and R H Clarke (London: Routledge)

[59] Bradshaw A M and Hamacher T 2013 Nuclear fusion and the helium supply problem *Fusion Eng. Design* **88** 2694–7

[60] Cai Z, Clarke R H, Glowacki B A and Nuttall W J *et al* 2010 Ongoing ascent to the helium production plateau—Insights from system dynamics *Resources Policy* **35** 77–89

[61] Sinars D B, Campbell E M, Cuneo M E, Jennings C A, Peterson K J and Sefkow A B 2016 The role of magnetized liner inertial fusion as a pathway to fusion energy *J. Fusion Energy* **35** 78–84

[62] Hockaday M *et al* 2011 *White Paper Compilation in Inertial Fusion Energy (IFE) Development (u)* (Los Alamos, NM: Los Alamos National Lab.(LANL))

[63] Pearson R J 2020 Towards commercial fusion: innovation, technology roadmapping for start-ups, and critical natural resource availability The Open University, UK https://doi.org/10.21954/ou.ro.0001187b

[64] Hoedl S 2019 A social license for nuclear technologies *Nuclear Non-Proliferation in International Law-Volume IV* ed J L Black-Branch and D Fleck (Dordrecht: Springer) pp 19–44

[65] Bradshaw A M, Hamacher T and Fischer U 2011 Is nuclear fusion a sustainable energy form? *Fusion Eng. Des.* **86** 2770–3

[66] Giegerich T, Battes K, Schwenzer J C and Day C 2019 Development of a viable route for lithium-6 supply of DEMO and future fusion power plants *Fusion Eng. Des.* **149** 111339

[67] Pearson R J, Bluck M J and Murphy S T 2017 A symbiotic approach to compact fission and fusion reactors *ANS Trans.* **117** 378–81

[68] McCarthy K *et al* 2002 Nonelectric applications of fusion *J. Fusion Energy* **21** 121–53

[69] Federici G *et al* 2016 Overview of the design approach and prioritization of R&D activities towards an EU DEMO *Fusion Eng. Des.* **109** 1464–74

Chapter 11

Commercial opportunities for nuclear fusion

William J Nuttall

11.1 Introduction and historical origins

This book considers the prospects for the commercialization of nuclear fusion technology in the second decade of the 21st century. This step has been a long time coming. The origins of nuclear fusion science can be traced to the 1930s and the work of Hans Bethe seeking to understand nuclear synthesis in stars, work for which he was in 1967 awarded the Nobel Prize in physics [1]. The early years of nuclear fusion research were linked to thermonuclear weapons development in the 1940s and 1950s. One of the first attempts to engage in civil nuclear fusion research was the Huemel Project in Argentina led by Ronald Richter, an emigre Austrian scientist, in the years after World War II [2]. The Argentinian civil power adventures led nowhere, but meanwhile fusion research was underway around the world in pursuit of a weapon that Hungarian–American physicist Edward Teller had christened the 'super'. In popular culture it became known as the 'Hydrogen-Bomb'. Despite the flurry of interest in the potential for nuclear fusion energy (for war or peace) the leading nuclear energy technology remained nuclear fission.

11.1.1 Fusion energy ambitions live in the shadow of nuclear fission success—1960s and 1970s

In order to understand early attitudes to the possibility of fusion energy it is helpful to consider attitudes to nuclear fission energy, or as it was known at the time: Atomic Energy. The earliest motivation for nuclear fission reactors in the late 1940s was related to the emerging science and engineering of nuclear weapons. In particular, there was a desire to increase understanding and capabilities required for the production of isotopes, particularly the plutonium-239 isotope associated with nuclear fission weapons development in the early Cold War. Such thinking motivated, for instance, the world's second reactor the X-10 Graphite Pile at Oak Ridge National Laboratory which first went critical on 4 November 1943. One of the less well-known stories of this reactor is that in 1948 a toy steam engine (a Jensen

doi:10.1088/978-0-7503-2719-0ch11

#50) was connected to the gas-cooled and graphite-moderated research and isotope reactor [3].The tiny steam engine turned a dynamo producing a miniscule, but non-zero, amount of electricity. The steam to electricity equipment was literally a toy, but the nuclear steam production system inside the reactor was not so trivial. It is quite remarkable that the staff's activities were entirely informal and initiated by the operator crew with some input from X-10 reactor engineers. As far as I'm aware, the project was never sanctioned by the leadership of the laboratory. It was a spontaneous initiative of the workforce. It came three years before the first official attempts to make nuclear electricity at the EBR-1 reactor in Idaho at what is now Idaho National Laboratory.

Why, you might ask are we concerning ourselves with these early stories concerning the weakly-regulated activities of early nuclear workers devoted not to matters of fusion, but of fission? The reason is that the examples of the X-10 and EBR-1 prove the early origins of the fascination of making electricity from atomic energy. It was a fascination that would endure for decades and that would infect decision making around anticipated nuclear fusion energy commercialization in ways that may have been unhelpful for a technology seeking a commercial application.

11.1.2 Atomic energy—the allure of electricity

Electricity has been the assumed pathway for the commercialization of fusion energy from the beginning, arguably from even before the beginning, as the early stories from fission reveal. Why did the first nuclear engineers have such a fascination with electricity as the peaceful application of their technology? There were in the post-war era plenty of process heat applications needing an affordable and reliable energy source, but in the 1940s and 1950s nuclear fission power (both gas cooled and light water technologies) could not achieve, both commercially and technically, temperatures for the most important industrial process heat applications. As regards the potential for the direct use of nuclear turbomachinery there was also a possible obstacle given the imposition of strict secrecy associated with early developments and ideas concerning nuclear naval propulsion for surface ships and submarines.

While one might suggest that the proponents of civil nuclear commercialization may have felt they were pushed towards nuclear electrical power production, I suggest that they went further and that they embraced it. In suggesting that visceral enthusiasm may have played a role I look to the historical context. A mid-career American nuclear engineer of 1950 might have been born in 1915. For those not growing up in the modern American cities spearheading the 20th century, the reality of their childhood would have been one of weak and rudimentary infrastructure greatly improved by the arrival of electricity in their formative years. The Rural Electrification Act was enacted on 20 May 1936 at a time when farms and rural communities could not be reached by the relatively low voltage US transmission infrastructure of the time. Rural electrification appeared to be impossible for the market to deliver and it was the progressive administration of President Franklin D Roosevelt that intervened to make the difference. Rural electrification was a cornerstone of Roosevelt's New Deal political platform to overcome the difficulties of the 1930s Great Depression. Such memories would lie in the background as the young engineers contemplated the societally beneficial application of the new nuclear technology.

The allure of atomic power was palpable—it even predated the demonstration of atomic weapons as revealed by the ideas circulating in science fiction in the late 1930s (figure 11.1).

After the war the allure of abundant cheap electricity from atomic energy did not diminish and in 1954 Levi Strauss Chairman of the US Atomic Energy Commission introduced the famous phrase 'too cheap to meter' when he said in a speech to science writers in New York City:

> It is not too much to expect that our children will enjoy in their homes electrical energy too cheap to meter; will know of great periodic regional famines in the world only as matters of history; will travel effortlessly over the seas and under them and through the air with a minimum of danger and at great speeds, and will experience a lifespan far longer than ours, as disease yields and man comes to understand what causes him to age. This is the forecast of an age of peace. [4]

For Strauss the peaceful benefits of nuclear energy would first and foremost be seen in the delivery of ultra-cheap electricity. From the perspective of the 21st century we can see that his *too-cheap-to-meter* prophecy was never achieved, but we can say that something did happen that might have surprised Strauss' audience even more. Today we find that telephone calls, and communications generally, are indeed too

Figure 11.1. The back cover of the October 1939 edition of science fiction 'pulp' magazine *Amazing Stories* featured a cutaway drawing of a future atomic power plant, as illustrated by Howard M Duffin.

cheap to meter in the internet age. Increasingly we pay simply for our connection, not the data delivered.

Not everyone shared Strauss' optimism regarding the cost of atomic energy. In the UK Sir John Cockcroft did rather better in his predictions, when he said in 1951:

> we do not expect to produce a cheaper source of power than that derived from coal—it is likely, in fact, to be somewhat more expensive. What we are aiming at is to increase the total power available [4].

It has been suggested that in his Utopian optimism Strauss was thinking not of the potential of nuclear fission power, but rather of the then more secret ideas of nuclear fusion energy of which he should have been aware through his leadership of the US Atomic Energy Commission. Such ideas remain, however, widely contested see e.g. the commentary by Thomas Wellock [5]. Going further, one can observe that Strauss did not actually make a link to nuclear energy (of any type) in his remarks but given his role and interests I think that connection, at least, can be safely inferred.

Whatever the anticipated costs of nuclear electricity and the balance of future effort between fission and fusion, it is clear that those concerned with the peaceful uses of atomic energy were focussed on the goal of large-scale electricity generation from the very beginning.

11.1.3 Fusion follows fission's footsteps

Even before the successful test of a nuclear fission weapon in the Manhattan Project's Trinity Test at the USAAF Alamogordo Bombing and Gunnery Range near Socorro, New Mexico on 16 July 1945, attention had turned to the energy to be gained from a thermonuclear weapon utilizing the principles of nuclear fusion. Those principles were demonstrated by the Mike device detonated by the Americans at Enewatak Atoll of the Marshall Islands in November 1952. Weaponization of the technology came later (the Mike device weighed 74 Tonnes), but Manhattan Project physicist Edward Teller's idea for the 'super' had now clearly been demonstrated. Human beings had demonstrated the release of fusion energy, but it was in the form of an explosion and not a controlled form suitable for commercial exploitation.

Meanwhile in the UK fusion efforts were underway entirely independent of American efforts, all wartime Manhattan project collaboration having been ended by the US following passage of the McMahon Act of 1946. For more on the story of Cold War UK nuclear weapons I recommend my book *Britain and the Bomb: Technology, Culture and the Cold War* [6]. The UK was active in both civil and military research, but things initially were not going well. The military efforts focussed on series of weapons tests known as Grapple and on 13 June 1957 the third shot of the UK Grapple series was fired. It yielded a disappointing 300 kT yield—less than might be achieved with a very large fission weapon. Britain was not yet a full member of the fusion club alongside the USA and USSR. They had been members of the club since November 1952 and November 1955 respectively.

On 8 November 1957 the British nuclear weapon test named 'Grapple X' achieved a Megaton yield, but at a huge cost in fissile material. The device had a thermonuclear (fusion) contribution, but in fusion terms it was not a great success. Britain still needed some good news. Only by late April 1958 and the 3 MT Grapple Y nuclear weapons test detonated near Christmas Island in the South Pacific Ocean was there a true British success sufficient to make it clear that the UK would be able to progress from a test to a viable H-Bomb weapon.

The weapons scientists from Aldermaston had finally achieved the key goal indicating British core capabilities for a thermonuclear weapon, but meanwhile across at the civil Harwell Laboratory things were going less well. Pressured by the Soviet success with the Sputnik 1 satellite in October 1957, and prior to the Grapple Y success, pressure was mounting at Harwell Laboratory for some good news. Work was underway on a plasma pinch controlled fusion energy machine known as ZETA. Scientists there were performing initial experiments aimed at achieving controlled nuclear fusion and they were observing an impressive release of high-energy neutrons. The news leaked to the press and a great British success was proclaimed (figure 11.2).

Soviet and American scientists were sceptical of the British claims and it was soon revealed that their scepticism was well placed—fusion had not been seen in south Oxfordshire. The neutrons were an artifact of some much more mundane physics occurring within the device. The failure was a huge embarrassment for the British government as they thought: *thank goodness Grapple Y had been a good result.*

In the decades after the end of World War II civil fusion research lagged both civil nuclear fission research and thermonuclear weapons research, but in the 21st century this is perhaps no longer the case. Civil fusion research has largely managed to avoid the controversies (waste, decommissioning, safety, and proliferation) associated with nuclear fission and the arms race of nuclear weapons is no longer obsessed with Megaton yields. Frankly, improvements in attack precision have eliminated the military case for the high yield 'H-Bomb' of the late 1950s. Today civil fusion research and development is strong and on the cusp of change with a shift from a world dominated by large international and collaborative public programmes (such as JET and ITER) to a new reality including high tech private sector start-ups such as Tokamak Energy and First Light Fusion in the UK, General Fusion in Canada and Commonwealth Fusion Systems in the US, among others. These companies are approaching fusion innovation in novel ways [7]. In the 21st century the commercial fusion landscape is crowded and fast moving, as this book explores. The world is very different than was seen in the chilliest years of the Cold War around 1960. Today the issues relate to innovation and commercialization not geopolitical technical superiority.

11.2 Civil fusion's first success

It should not be forgotten that fusion science and technology has already been commercialized, but of course not for electricity generation. The story of the 'Fusor' is informative in that regard. One key lesson from the story of the Fusor is the notion

DAILY SKETCH 2½d

SATURDAY, JANUARY 25, 1958

OUR SCIENTISTS SPUTNIK THE RUSSIANS—FOR PEACE

A SUN OF OUR OWN!

AND IT'S MADE IN BRITAIN

P-S-S-T!

By PETER STEWART

HARWELL, BERKS, Friday

BRITISH scientists revealed their stupendous ZETA secret officially to-night. They have built a man-made sun on earth.

It is a peaceful, safe sun which promises limitless cheap power for all mankind.

THE HEAT FROM 10 TONS OF COAL FOR 2s

MUCH HOTTER

SO SHY

P-S-S-T!

WHAT'S THE WEATHER?

Owbridge WEATHER!

Owbridge's

LUNG TONIC AND PASTILLES

FOR COUGHS AND COLDS

Figure 11.2. Cover of now defunct Daily Sketch 25 January 1958. (Source: ITER (https://www.iter.org/newsline/-/2905).

that a commercial fusion device needs to achieve energy 'break even' in order to be commercially successful. The Fusor appeared to have no way forward, but such a conclusion can be erroneous. The logic was predicated on the notion that the commercial product would be energy in some form (probably electricity), but fusion commercialisation need not be based on energy sales at all. Fusion neutrons could be important for some potential customers. Neutrons are sufficiently valuable to some users that they will be willing to pay even if very large amounts of electrical energy must be put into the machine to make them. Indeed, the pathway to Fusor commercialization has followed that path. The story reminds us that the future of fusion does not necessarily rely on finding a path to abundant cheap electricity, but rather on finding a pathway to commercialization by any means.

One of the most attractive opportunities for fusion commercialization relates to the commercial uses of 14 MeV neutrons as produced by D–T fusion. Relevant Industrial Sectors that might make commercial use of such neutrons include:

- Direct medical uses—such as perhaps Boron Capture Therapy
- Diagnostic and Therapeutic Medical Isotope Production—e.g. Technetium-99m
- Technical support to civil nuclear energy, including the fission fuel cycle.

I am aware of at least three companies focussed on the commercial use of fusor technology as a neutron source. The three companies are:

- NSD Gradel of Luxembourg (http://www.nsd-fusion.com/)

NSD advertises a range of small fusor-based neutron generators in its marketing brochure [8]. For their NSD-350–24-C Neutron Generator the company advises that in D–D mode a neutron yield greater than 10^8 per second can be achieved.

- Phoenix LLC of Monona Wisconsin, USA (https://phoenixwi.com/)

Phoenix claims to build the strongest compact neutron generators in the world [9]. The company's website asserts that they use a:

> 'miniature particle accelerator and an ion beam to cause a fusion reaction and produce tens of trillions of neutrons per second, Phoenix's high-yield, high-flux neutron generators provide a compact, accessible, clean alternative to reactor facilities for non-destructive testing and other industrial applications.'

- Daimler-Chrysler Aerospace

Daimler-Chrysler Aerospace was the first to market with a commercial fusion proposition with its 'FusionStar' technology in the 1990s. The endeavour was not a commercial success and the products were removed from the market shortly thereafter [10].

Fusor research continues around the world. A large amount of effort went into the related concept of magnetically confined fusor technology relying on the notion of 'polywell' or 'Wiffle Ball' confinement. While there was significant effort over several decades by 2019 a consensus was emerging that any such plasma will always preferentially escape along paths that must exist where the magnetic fields are forced together. Several companies including EMC2 Fusion Development Corporation devoted much effort to the concept (http://www.emc2fusion.org/), but at the time of writing a commercial product has not yet been achieved.

Meanwhile research has continued into traditional electrostatic fusors but with the possibility of time-varying the potential applied to the grid—such as via harmonic oscillations. Such work has given rise to the POPS concept Los Alamos National Laboratory in NM USA. 'POPS' stands for: Periodically Oscillating Plasma Sphere. It would appear that dynamic fusor technology could yield improvements over traditional devices especially if it can be combined with advanced grid technologies, such as might be possible using carbon nanoforms related to the science of graphene.

Lastly it is appropriate to pause and reflect that fusors are sufficiently simple that they have given rise to a global community of hobbyists making and testing their own small devices [11]. One such community is the High Energy Amateur Science Group in Virginia, USA.

To close our short digression into the science and technology of fusors it is appropriate to remember the words of Paul Schatzkin: *Fusion is not* 20 *years in the future. It's* 50 *years in the past—and we missed it.* To which I would only add, the scientific community may have turned away for some time, but to this author's impression the fusor story is definitely a key component in understanding the commercial potential for fusion that lies ahead. Not all fusion is difficult.

11.3 The painful story of 'cold fusion'

Any commentary on the commercialization of fusion technology should not ignore the painful story of Cold Fusion. In 1989 the then eminent electrochemist Martin Fleischmann and his former student Stanley Pons shocked the world when they announced via a press conference that they appeared to have produced nuclear fusion energy from a small electrochemical cell. This awoke interest among potential investors in a new, cheap and abundant source of energy. Several start-up companies were formed in the hope of commercializing the new science—indeed, noting the focus of this book, it has to be said that the first Japanese commercial fusion enterprise was a Cold Fusion company. In the US, meanwhile Energetics had an active Cold Fusion programme as did several other companies. Meanwhile major laboratories around the world were seeking to replicate the initial reports from Fleischmann and Pons. The problem was that the major laboratories, including Harwell in the UK, could not reproduce the results and furthermore no-one was seeing any neutrons which should be expected if the posited nuclear fusion were indeed occurring. Seemingly no-one anywhere was making tangible progress.

One by one the small start-ups failed, larger corporations turned away, including Toyota which went so far as to set Fleischmann and Pons up in a laboratory in France. In the US Energetics eventually gave up their Cold Fusion interests and the universities gradually aligned behind the view that in Cold Fusion there was nothing to see.

The Cold Fusion story is well known by everyone associated with today's fusion challenges and several lessons can be drawn from it. Arguably it feeds into a distrust of fusion commercial ventures (of any type) in corners of some of the large publicly-funded laboratories. It also acts as a warning to those seeking to promote unorthodox approaches to fusion in either the public or the private sector. A key lesson is: do not let the publicity get ahead of the science. One of the long-standing concerns around Fleischmann and Pons is that they released their original results in a press-conference and not in the prestigious peer reviewed journal, *Nature*, as they had intended.

It is sometimes said that the flurry of interest in Cold Fusion in 1989 was a consequence of the very real science of high temperature superconductivity revealed to the world in 1986 by Alex Mueller and Johannes Georg Bednorz and which had seen them win the Nobel Prize just one year later. They had published their work in the scientific literature and moreover they had chosen a modest and tentative title: *Possible high Tc superconductivity in the Ba–La–Cu–O system*. For Fleischmann and Pons things were to be very different.

In 2020 the global community of Cold Fusion proponents has dwindled, but among the remaining researchers a few stand out—one such researcher is Dr Peter Hagelstein of MIT. He features strongly in a *Washington Post* article from 2004 entitled: Warming up to Cold Fusion [12]. Hagelstein rebuts many of the criticisms of Cold Fusion science. In so doing he adopts a modest and good-humoured approach, despite the great setbacks to his career that have no doubt flowed from his advocacy of something that is conventionally regarded as pure bunkum.

Hagelstein and the few remaining researchers talk these-days about Low Energy Nuclear Reactions rather than 'Cold Fusion'. The renaming appears to reflect a sense that what is happening is probably not fusion in any conventional sense, but the advocates remain firm in their sense that something important can happen in these simple electrochemical cells, if you do the experiments carefully enough. After more than 30 years, however, no useful energy has ever been extracted from a Cold Fusion cell. The early hopes are clearly unfulfilled.

As one of a set of authors focused on commercialization, I feel compelled to pause and mention one claim from the Cold Fusion community—that they have reproducibly observed the correlated production of helium, heat and x-rays in D–D devices. If that is true that would be extremely interesting, but even more interesting is the suggestion that it might be possible, by similar means, to make extremely valuable helium-3 using inputs of hydrogen and deuterium. If Cold Fusion is not fusion, but is something that could make helium-3, then the world would be interested, indeed the fusion world would be interested as the helium-3 fusing with helium-3 reaction is a relatively accessible (low energy) reaction that should be easier to contain the more conventional D–T approaches. The problem has been historically that helium-3 is only available in minute quantities. It is one of the most expensive substances known to man. In the absence of a cheap abundant source it is not a fuel for affordable fusion. As regards the Cold Fusion manufacture of helium-3, the fact is that such ideas are now more than ten years old and to this author's impression nothing has happened, to demonstrate feasibility. This rather implies that Cold Fusion has no future as a fusion fuel maker as well as being dead, as orthodox thinking already believed.

As for the risk that the memory of Cold Fusion could adversely affect the prospects for the current wave of fusion commercialization, well the shock and disappointment of the Cold Fusion story are probably now long enough ago to have been forgotten—time is a great healer.

11.4 Looking ahead: fusion's potential commercial attributes

It is perhaps natural to seek to compare the commercialization to the now well-established paradigm of commercial fission-based nuclear power. However, it is also important to consider the ways in which fusion presents a more attractive set of characteristics than the more established fission alternative. Two major societal concerns relating to nuclear fission power are first the risks of severe nuclear accidents as seen at Chernobyl, Ukraine in 1986 and at the Fukushima-Daiichi plant in Japan in 2011; and second the problems associated with the safe management of radioactive wastes for thousands of years.

In response to concern about large scale accident risk, the fusion community is right to assert that the risk of a major accident requiring an off-site response is vanishingly small, in part because of the very low levels of stored energy in the system. A nuclear fission plant is typically fuelled every 18 months and in principle that provides a rough guide as to how much energy is available for release in a worst-case accident scenario. In contrast a fusion power plant (whether magnetically of inertially confined—see chapter 1) is a device that is refuelled continuously or at rapid intervals. There is simply not the fuel available for anything truly horrendous to happen. In the case of a fusion tokamak the greatest stored energy is likely to be in the superconducting large magnets—but even if that were all to be released in a sudden event it would do no more than destroy the tokamak with little chance that any debris would escape the building. There are of course risks of dust and fire but these can be expected to be less radiologically problematic than any equivalent scenario in a nuclear fission plant.

As regards radioactive wastes. For conventional D–T nuclear fusion (see chapter 1) the product of the fundamental nuclear reaction is the inert and stable gas helium-4. There is no problem concerning radioactive spent fuel as is seen in nuclear fission power industry. Fusion, like fission, must plan for plant decommissioning at the end of life and it is here that a radioactive waste challenge can be expected. Such concerns have motivated the European Union's efforts in fusion delivered under the auspices of the Euratom Treaty. The EU has favoured efforts towards concepts in which long-term radioactive wastes are avoided (where we take long-term to mean of concern for more than 100 years). This has led to the development of special low activation structural materials, such as Eurofer martensitic steel. We must pause to note that while one can only admire the materials science and engineering achievements of the researchers involved, these developments were not necessarily wise in terms of meeting the goal of rapid fusion commercialization, as they do not assist in meeting the challenge of speed and they risk being motivated by problems that perhaps do not even exist—concerning the public acceptance of nuclear fusion energy. Perhaps more progress would have been made if the focus had been able to build a fusion power plant as quickly as possible in a country that might want one, rather than focussing on the goal of construction in a place where the technology could be expected to meet with some scepticism or even hostility. The task of addressing the concerns of the most opposed can lead to delay and additional technical challenges beyond the issues of fusion demonstration. For the European

Union, a long-standing supporter of civilian fusion energy, the Euratom Treaty and its requirements for member state unanimity concerning strategy might be part of the story. Territories without such a political and constitutional framework might be able to make more rapid fusion engineering progress, but some might add—at what cost?

It is clear that the fusion community can make a valid and compelling case that their technology might one day generate electricity more safely and with a lower radioactive waste burden than fission power does today. If to this can be added success in achieving a lower cost than renewables with power storage, then fusion power could indeed be an important part of our electricity future. Concerning a beneficial fusion attribute, safety, in comparison to fission power generation, a fusion power plant has no post-shutdown decay heat management challenge. It was fission-product decay heat management failure that lay at the heart of the accidents at Three Mile Island Pennsylvania in 1979 and Fukushima-Daiichi, Japan in 2011. Also, fusion is entirely spared the related, and more general, concerns around loss-of-coolant accident risks in fission power plants. In essence there is no large-scale heat source in a fusion power plant that requires oversight and control after shutdown.

It is sometimes posited that nuclear fusion energy presents a substantially lower nuclear weapons non-proliferation risk than fission power. Such statements are easily made but are harder to justify in a robust way. Consideration of such matters would need to go beyond consideration of the physics of fertile to fissile actinide transmutation in the presence of 14 MeV fusion neutrons to consider issues including: technical difficulty, proliferation cost, proliferation time and the risk of detection [13] I am not saying that D–T devices represent a worsening of proliferation risk when compared to fission alternatives, I am merely saying that —in my opinion—the case is hard to assess.

11.5 Fusion can do better than electricity generation

We have seen earlier that fusion has tended to follow fission towards a goal of large-scale electricity generation in a spirit of the peaceful use of nuclear energy. The merits of fission in that regard can be debated elsewhere, but in this book we are focussed on the commercialization of fusion energy, a goal that has still not been achieved nearly 70 years after the detonation of the Mike device in 1952.

I suggest that fusion commercialization has not been achieved as quickly as it might have been because electricity was not the best first goal for the early fusion developers. The stories of fission and fusion have fundamentally different beginnings. The science and basic engineering of fission were first demonstrated at-scale in a controlled, stable and continuous state via the CP-1 reactor built by Enrico Fermi's team in a squash court at the University of Chicago. The reactor first went critical in December 1942. The first detonation of a fission weapon came 31 months later, in July 1945, with the Trinity test in New Mexico. Fusion has no parallel sequence. For fission the possibility of peaceful application in the form of the stable baseload production of electricity was manifest from the start. It was clear from 1942 that it would be possible to maintain a fission reactor in a stable state. In this way it was clear from the outset that one might generate commercial power using the heat generated by the nuclear fission process. That commercial potential was clear, before the demonstration of the enormous amounts of energy available that came with the later Trinity Atomic Bomb test. For fusion controlled stable conditions have still not been achieved more than 70 years after mankind first uncontrolled released fusion energy. In some sense fusion does not have an equivalent of the working reactor assembled in a university squash court.

The origins of the electricity business involve the early use of hydro-power, but the combustion of fossil fuels (originally coal) to generate steam to turn a generating turbine soon followed. The Edison Electric Light Station was constructed at 57 Holborn Viaduct in London in 1882. For more than a century fossil fuel combustion lay at the heart of the electricity sector and defined its economics. Capital costs were modest, wastes and emissions were essentially ignored and only two prices really mattered—the price of fuel (costs) and the price of electricity (revenues). In the late 20th century things started to change with the emergence of nuclear fission power and later non-dispatchable renewables (wind and solar). These new methods typically had very high capital costs and very low, or zero, 'fuel' costs. As such, these new technologies had very low marginal costs of generation. The costs they represented to the operator were essentially the same if they ran and if they did not. Intermittent and varying renewables such as wind and solar energy have grown enormously in importance, but they have tended to be supported by subsidies causing them to be a more attractive proposition than they might otherwise have been. Today, however the need for subsidy is now greatly diminished. While nuclear fission technology has also benefitted from subsidy (especially in research and development) that technology tended in the years after 1990 to present itself as a commercial proposition with an economic basis very different from that of fossil fuel combustion. While it is technically possible to operate modern nuclear fission plants variably, such as in load-following mode, typically such a strategy makes no economic sense [14]. The point is that as long as power prices are not actually negative in the market, and while the marginal costs of generation are effectively zero, it makes no sense to turn off the plant when demand, and hence price, is low. Turning off the plant is turning off a small amount of revenue, but it does nothing to reduce costs. Why forgo revenue, even a small amount of revenue? The economics of electricity dictate that conventional fission-based nuclear energy is naturally a baseload technology. Any technical considerations merely amplify the case for a baseload role, as all machines run better when operating continuously at the sweet spot of the design parameters. Such engineering realities underpin the logic of a hybrid car (e.g. the Toyota Prius).

I suggest that the commercial proposition for nuclear fusion electricity generation is exactly equivalent to the fission proposition. The technology will be high capital cost and with very low marginal costs of generation. Fusion power must surely also operate as a baseload generator for economic reasons. I concede the market of the future might be different than today, but I remind readers that if they are thinking a low-carbon subsidy will come to the rescue of fusion electricity generation, they must recognize that the electricity system will have been entirely decarbonized by other means more than 20 years before fusion might first generate commercial power. Fusion advocates should start to turn away from addressing a problem (the challenge of low-carbon electricity) that may cease to exist in the relatively near future, and importantly long before affordable and reliable fusion electricity arrives. Importantly fusion can also do good in ways unrelated to electricity generation and indeed, as a consequence of the climate challenge, it must.

As noted earlier—the sustained and controlled production of net-fusion energy has never been demonstrated, but the explosive uncontrolled release of fusion energy has been in the hands of mankind for nearly 70 years. Why does continuous fusion energy demonstration elude us? In part it is because at their heart all fusion propositions are pulsed devices. Inertial confinement fusion (ICF) has direct parallels with the science of thermonuclear weapons and hence it is unsurprising to argue that the proposition there is for pulsed operation. Typically, continuous baseload power production would be achieved by the ignition of a series of ICF capsules like bullets in a machine gun belt.

Magnetic confinement fusion (MCF) has no direct links to nuclear weapons science, but it too is essentially a pulsed technology. Fundamentally an MCF tokamak is a transformer. The ionized plasma of, for instance, deuterium and tritium plays the role of the conductor in the secondary winding of a transformer. The primary winding might, for example in a 'spherical' tokamak (see chapter 2), be a key part of the central column running down the middle of the tokamak torus. In modern tokamaks energy is deposited in the plasma by several means, such as neutral beam injection or via high frequency electromagnetic waves—akin to the science of a microwave oven. The original idea, however, is to heat the plasma via transformer action, not in alternating current mode as is usual for a transformer, but in a single sweep in voltage from strongly positive to strongly negative applied to the primary. This will, by transformer action, drive rotation (actually, counter rotation of ions and electrons) in the plasma. The plasma is similar to a viscous fluid and the sudden drive to rotation will generate resistive/frictional heating. The plasma starts to rotate and warm, as a result of that initiating sweep of voltage applied to the primary. The key consideration, however, is that it is a single event—it is not maintained, and it is not easily repeated. It is in essence a pulse, albeit of much longer duration than what is seen in ICF approaches. Once hot and rotating a tokamak plasma can be sustained in both its temperature and rotation, but this is not easy.

Fission power reactors are brought to criticality through the deliberate and short-term injection of neutrons. This is typically provided by a start-up neutron source based on Californium-252 as a spontaneous neutron emitter or from a strong alpha emitter in the presence of beryllium. I would suggest that this initiating event in a fission power plant is not equivalent to the initiation of a tokamak achieved by transformer action. In the former case if one did not have such an initiating neutron source the reactor could in principle spring into life by itself (initiated perhaps by cosmic rays). This is not done because it brings in easily avoided safety risks. The start-up neutron source in fission makes life easier, it is not fundamental to the proposition of fission power production. Transformer action, meanwhile, is fundamental to all tokamaks and it is this reality that leads me to suggest that even tokamaks are, in essence, pulsed machines. I am not saying it is impossible to extend the pulse for baseload operations, my point is simply that it needs to be done. It remains to be seen how long a plasma can be easily sustained and indeed plasmas have now been run for days, but if fusion is to match fission as a baseload power technology 18 months of continuous full power operation is needed with then an 8-week shutdown for scheduled maintenance before another 18 month run. That is what baseload low carbon power looks like today, 40 years before fusion seeks to do better.

Even if fusion could achieve economically viable and technically sustainable baseload power generation an even more serious obstacle to commercial success comes into view. Investors in the electricity business, like investors everywhere, are concerned fundamentally by two things—expected return on investment and economic risks. Typically, the bulk of concern relates to risk awareness, management and mitigation. Fusion electricity generation is very different from fission power in this regard. If a large fission power plant trips off-line and the reactor shuts down suddenly in a 'SCRAM', then typically an extended period is required for reset before restart. In part this is needed to allow for the xenon-135 fission product in the fuel to decay away. Xenon-135 has a 9.2 h half-life and it is a strong neutron 'poison' meaning that it soaks up neutrons making reactor criticality harder to achieve. While in principle one could restart quickly it is easy, and prudent, usually to wait. Once ready to return to operations, all that is required is the use of the start-up neutron source and one is back in business. Little or no demand is placed on the electricity grid. MCF tokamak fusion would be very different.

A large MCF tokamak power plant might trip-out unexpectedly. As with a fission plant shutdown, at that point the grid may lose more than 1 GW of expected baseload power. At the fusion power plant checks will be made and systems reset. After, say 45 min, the operators of the fusion plant might contact the electricity system operator to say that they are ready to return to generation. At this stage there is an important difference between a fusion and a fission power plant, and the difference is not good for the case for fusion. One might think the fusion power plant operator's communication that the plant is ready to return to generation would be received with delight at grid control room given the need to cope with 1 GW less generation than had been expected. That would, however, probably not be the case. I posit that the grid control room would not receive the communication from the fusion plant with enthusiasm, because a key part of the message must be that the fusion plant needs to withdraw a few hundred MW of electricity from the grid for several seconds, perhaps even minutes in order to be ready to fire again. That will be electricity that the grid can ill afford to make available at a highly stressed time.

It is sometimes argued that tokamak fusion avoids the need for decay heat management (as is required in a fission power plant) and hence this avoids the need for tens of MW of on-site diesel generators as a last backstop in the case of grid connection failure. I suggest that such claims risk being advanced too strongly given the pressures that a tokamak power station operator would be under to be able to restart without requiring tens of Megawatts from the grid system. The restart power demand problem for tokamaks can be fixed, but it requires infrastructure investment that worsens the problem that the technology is too capital intensive. Economic modelling can measure and assess these issues of technical reliability and market risk [15, 16]. Difficult trade-offs between cost and economic risk may be required.

11.6 The importance of net-zero and deep decarbonization

In the preceding section I have laid out a case suggesting that fusion commercialization (especially tokamak MCF commercialization) might not be most easily advanced via the generation and sale of electricity. That does not mean that I see little prospect for the successful commercialization of fusion energy. Nor does it mean that I believe that fusion generated electricity will not find its place. I see every prospect in both cases. My proposition is that commercial might emerge more quickly than has previously been imagined, but that commercial success is not likely to be seen first in electricity.

I have also argued earlier that fusion might come too late to help with the decarbonization of the electricity sector, but it will not come too late for the decarbonization of energy and industry more generally. Fusion energy most definitely has a useful role to play. Thus far the nuclear industry, fission dominated but also including fusion, has tended to see its role as decarbonizing electricity—the easiest and cheapest to achieve part of the global decarbonization challenge. I suggest that the nuclear industry, and most especially the emerging fusion industry should stop focussing on the first 20% of the decarbonization challenge and instead orient themselves to address the last 20%. The last 20% of decarbonization meets with the growing sense that the world needs to aim for 'net-zero' emissions by 2050, and indeed this is now UK Government policy. Going in this way for 'deep decarbonization' will take nuclear energy into the hardest to decarbonize sectors. These challenges include issues in industrial process engineering (process emissions and process heat) and transport and mobility (especially shipping and aviation). Some parts of these sectors are hard to address via electrification and hydrogen has been posited as an alternative low-carbon energy carrier to electricity or as an alternative reducing agent to coking coal in iron and steel manufacturing. Nuclear energy, including fusion energy, may have a beneficial role to play in the large scale low-carbon production of hydrogen for industrial and mobility applications. With A T Bakenne I have considered the prospects for hydrogen energy in another work [17]. We shall return to consideration of hydrogen later, but let us first step back and consider some other opportunities for the successful commercialization of fusion science and technology.

11.7 Non-electricity commercialization

In section 11.5 I argued against the likelihood that the first successful commercialization of MCF tokamak technology would be in the arena of electricity generation. To have a more optimistic sense of the commercial opportunities ahead for fusion it is useful to take a step back and consider possible commercial products. These 'products' could be small scale fusion devices sold directly to end users, or a start-up fusion company might seek to sell a service providing 14 MeV neutrons from D–T fusion, or an industrial concern might seek to sell fusion process heat, such as from the blanket of a tokamak.

11.7.1 Process heat applications

While the production of fusion neutrons represents the simplest commercial service that might be envisaged, the next most ambitious goal is also significantly more attractive than the challenges of electricity generation. The intermediate goal relates to industrial process heat applications. This links back to the challenge presented by the hard to decarbonize sectors considered in section 11.6. One particular process heat application concerns the rapidly emerging hydrogen economy of the 2020s and 2030s.

More than 15 years ago I was conscious that fusion would not easily achieve commercial success via electricity generation. Such thinking led me in 2004 to write an opinion piece in *The Engineer* magazine entitled *Fusion Should Put Its Energy into Oil* [18]. I was attracted to the notion that fusion would be a safe and low-carbon route to hydrogen manufacture via high temperature thermochemistry. I was imagining a tokamak and associated infrastructure producing liquid hydrogen for international shipment. The activity would be managed by companies that had once been thought of as international oil companies and the enterprise would be reminiscent of aspects of today's liquefied natural gas industry. The whole endeavour would represent a strategic next step for today's international oil companies. They have the right culture and risk appetite to deal with the uncertainties and costs associated with fusion and in a low-carbon world they need to change. As time moved on I became more conscious that the fundamental fuels for D–T fusion (essentially lithium and deuterium) have good natural abundance and hence probable fuel security, notwithstanding the pressures on lithium supply caused by the use of the metal in modern batteries. Fuel may be secure and abundant, but the large superconducting tokamaks require very large amounts of helium—a potentially unsustainable by-product of today's natural gas economy. This concern was first explained to me by Richard Clarke and it was later explored by our PhD student Zhiming Cai. Large-scale and long-term helium supply is a concern for those planning superconducting tokamak-based fusion energy solutions [19, 20]. A breakthrough came when Bartek Glowacki at Cambridge University pointed out that liquid hydrogen could represent an excellent coolant for some high temperature superconductors that might be useful in tokamak design. The *Fusion Island* concept was born [21, 22].

Figure 11.3 highlights the key elements of the 2005 Fusion Island concept. The fundamental premise is the production of 20 K cryogenic liquid hydrogen by a facility wholly owned and controlled by an international oil company. It would be a low-carbon liquid fuel factory to replace a crude oil reserve in the asset portfolio of an oil company. The cryogenic liquid hydrogen is the product sold to global markets in a manner similar to liquefied natural gas, but the use of cold hydrogen also avoids the need for large amounts of expensive and increasingly hard to source helium for

Figure 11.3. Artist's impression of Fusion Island (Source: *The Engineer* [21] with kind permission).

magnet cooling. The system has no requirement for continuous reliability in that the product is collected from the island only intermittently from a large scale-storage capacity.

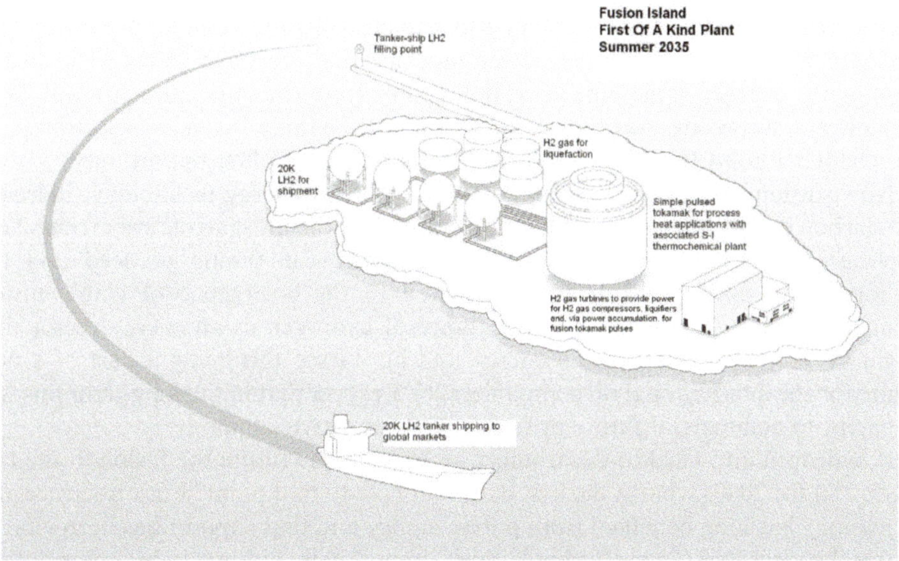

The 2005 Fusion Island concept pointed the way to future oil company engagement with fusion technology. In 2018 the Italian oil company ENI partnered with the Massachusetts Institute of Technology and MIT spin-out Commonwealth Fusion Systems with a view to finding a possible 'game-changer' in the energy sector [23]. Meanwhile British oil company BP has hosted discussions surrounding the potential future importance of fusion [24].

I have long seen the best way ahead for fusion as an energy technology to be via the emergence of a hydrogen economy. Indeed I co-hosted a UK Energy Research Centre conference to that effect in 2007 [25]. In the years since I have come to understand better the pathways to a hydrogen-based energy system. The Fusion Island concept, as initially described, relied on thermochemical hydrogen production from water. Over the last 15 years progress in such technology has been slow. The sulfur–iodine cycle proposed for Fusion Island remains extremely challenging owing to the high temperatures required (roughly 800 °C) and the need to use some rather nasty concentrated acids. Over the years there has been much discussion of green hydrogen—i.e. hydrogen produced from otherwise surplus renewable electricity. I however have become most interested in the benefits obtainable from natural gas sourced hydrogen linking to carbon capture utilization and storage (CCUS)—this has been termed 'blue hydrogen'.

I now see the opportunity for an updated conception of Fusion Island linking low-carbon fusion-based process heat with natural gas methane reformer-based hydrogen production and CCUS. Together with Co-Editor David Webbe-Wood I hope to give this matter further consideration in the coming months.

11.8 Conclusions

In this chapter I have pointed to the curious fascination that the nuclear community has had with nuclear electricity generation. The fascination is older than the first nuclear reactor and it is not surprising that nuclear fission technology went in that direction giving us the nuclear power industry that we have today. The fusion community emerged some time later, but coming from the same nuclear family, it is perhaps not surprising that they sought to join the family business—electricity. In this chapter I point to the risks inherent in that choice of first opportunity.

Now, fusion, of course, has the potential to be an energy technology, indeed a low-carbon energy technology. I have elaborated on the most attractive process heat applications in hard-to-decarbonize sectors where the timing of need and the willingness to pay are likely to match well to the emergence of viable fusion technologies. I have stressed the importance of hydrogen as an energy vector that might be produced using fusion energy and how I see this being a part of a new future for the international oil companies [26]. I expect that in the long-term this will represent to commercial future of fusion as an energy technology.

If hydrogen and hard-to-decarbonize sectors are the future for fusion in the late 2030s and the 2040s what is the best path to follow to that point? Fusion science and technology has long benefited from public money and that support has been vital to enable the ideas laid out in this book. Such support will continue to be important for many years to come, but I suggest that while such support is necessary, it is not sufficient. There is a need for private sector fusion with its different culture and attitudes to risk [7]. I posit that success is best not achieved by a sense that fusion commercialization lies in the future, it must increasingly focus on the fact that fusion commercial success exists already. The clearest example of such success relates to technologies in the family of technologies associated with the Fusor concept.

As suggested above, the most immediate pathway to the commercialization of services delivered by nuclear fusion physics is the production of neutrons for industrial and medical purposes. Businesses built on such technologies, such as Phoenix in Monona, Wisconsin exist today and they do not produce electricity—interestingly the key relationship to electricity is that they tend to use it, not make it.

In conclusion, I posit that perhaps too few people in the fusion science and technology arena have paused to think deeply about the nature of the electricity business and their sector's possible interactions with it. It is perhaps too easy to think that the looming abundance of cheap low carbon electricity is a threat to the case for commercial fusion as a low-carbon energy technology. My suggestion, however, is that the availability, even if only at certain times, of cheap (and clean) electricity is a huge opportunity for the short-term emergence of novel fusion-based innovations. I suggest that any small high tech company contemplating the improved use of, for example, accelerated deuterons, larger-scale fusor devices, or any one of a myriad of other clever physics ideas should rejoice in the fact that their electricity bill, and their carbon footprint, should not be a problem as they seek to mitigate disease, enhance industrial safety or improve homeland security via the neutrons or process heat that they hope to offer for sale. Some of these opportunities are near term and in areas where few affordable, clean, and reliable options currently exist. While the future for fusion electricity remains some way away, the future for commercial fusion may be with us already.

Acknowledgements

I am enormously indebted to Bartek Glowacki and Richard Clarke for their assistance and friendship over many years. Together we were at the heart of the development of the 'Fusion Island' concept. While we didn't get everything right, I am pleased that to some extent at least we did see the future that is now increasingly visible. We were helped along the way by very many wise counsellors too numerous to mention, but I would like to stress the importance of Joe Minervini, Leslie Bromberg and Bruno Coppi at MIT who kindly guided us in our understanding of tokamak technologies. I am also most grateful to *The Engineer* magazine and to *Nuclear Engineering International* for giving space in their publications for ideas that were, at the time, far from the mainstream. Finally, I would like to thank Richard Pearson and the late Stephen Smith for comments and advice concerning Fusor technology and many other things. Of course, everything presented here represents my own thoughts and opinions and should not necessarily be linked to those that have helped me along the way.

References

[1] Nobel Media AB 2020 The Nobel Prize Organization *Hans Bethe Facts* (https://nobelprize.org/prizes/physics/1967/bethe/biographical/)

[2] Fantoni E 2011 *Nuclear island: The secret post-WWII mega lab investigated Wired Magazine* (https://wired.co.uk/article/nuclear-island)

[3] Youtube video entitled: *World's first nuclear power generated electricity from Jensen #50 on the X 10 Graphite Reactor 1948* uploaded on 25 January 2015 by user Reddrryder (https://youtube.com/watch?v=FzAjO5_SyTU)
[4] Brown M J Canadian nuclear society *Too Cheap to Meter?* (https://cns-snc.ca/media/media/toocheap/toocheap.html)
[5] Wellock T 2016 NRC Blog *'Too cheap to meter': A History of the Phrase* (https://public-blog.nrc-gateway.gov/2016/06/03/too-cheap-to-meter-a-history-of-the-phrase/)
[6] Nuttall W J 2019 *Britain and the Bomb: Technology, Culture and the Cold War* (Dunbeath Scotland: Whittles Publishing)
[7] Pearson R J, Costley A E, Phaal R and Nuttall W J 2020 Technology roadmapping for mission-led agile hardware development: a case study of a commercial fusion energy start-up *Technol. Forecast. Soc. Change* **158** 120064
[8] GRADEL s.à r.l. (undated) NSD gradel fusion neutron generators Marketing Brochure (http://nsd-fusion.com/template/userfiles/files/broschuere.pdf)
[9] Phoenix LLC Corporate Website (https://phoenixwi.com/neutron-generators/)
[10] Sved J 1999 The first IEC fusion industrial neutron generator and developments *AIP Conf. Proc.* **475** 704
[11] Dunn W 2018 *The Amateur scientists tackling the global energy crisis* (New Statesman) (https://newstatesman.com/politics/energy/2018/12/amateur-scientists-tackling-global-energy-crisis)
[12] Weinberger S 2004 Warming up to cold fusion *The Washington Post* (https://washingtonpost.com/wp-dyn/articles/A54964-2004Nov16.html)
[13] Ashley S F *et al* 2014 Fuel cycle modelling of open cycle thorium-fuelled nuclear energy systems *Ann. Nucl. Energy* **69** 314–30
[14] Pouret L 2009 Nigel Buttery and William Nuttall, is nuclear power inflexible? *Nuclear Future* **5** 333–41 and pp. 343-344
[15] Steer S J, Cardin M-A, Nuttall W J, Parks G T and Goncalves L V N 2012 Minimising the economic cost and risk to accelerator-driven subcritical reactor technology: the case of designing for flexibility: Part 1 *Nucl. Eng. Des.* **243** 135–47
[16] Cardin M-A, Steer S J, Nuttall W J, Parks G T, Goncalves L V N and De Neufville R 2012 Minimizing the economic cost and risk to accelerator-driven subcritical reactor technology. Part 2: the case of designing for flexibility *Nucl. Eng. Des.* **243** 120–34
[17] Nuttall W J and Bakenne A T 2019 *Fossil fuel hydrogen-technical, Economic and Environmental Potential* (Berlin: Springer)
[18] Nuttall W J 2004 Fusion should put its energy into oil *The Engineer* 18 (https://theengineer.co.uk/fusion-should-put-its-energy-into-oil/ accessed 22 September 2020)
[19] Nuttall W J, Clarke R H and Glowacki B A (ed) 2012 *Future of Helium as a Natural Resource* (London: Routledge)
[20] Zhiming Cai R, Clarke B A, Glowacki W J, Nuttall and Ward N 2010 Ongoing ascent to the helium production plateau—insights from system dynamics *Resour. Policy* **35** 77–89
[21] Nuttall W J, Glowacki B A and Clarke R 2005 A trip to 'Fusion Island' *The Engineer* 16–8
[22] Nuttall W J and Glowacki B A 2008 Viewpoint: fusion island *Nucl. Eng. Int.* **53** 38–41
[23] ENI Press Release 2018 (https://eni.com/en-IT/media/press-release/2018/03/eni-joins-mit-and-cfs-to-advanced-fusion-energy.html)
[24] Lamont A(Host) 2020 BP technology outlook fusion podcast fusion *Topic 12 podcast* (https://bp.com/en/global/corporate/what-we-do/technology-at-bp/technology-outlook.html)

[25] O'Brien M 2007 UK energy research centre meeting report, *Hydrogen Production: A Role for Fusion?* (https://ukerc.rl.ac.uk/UCAT/PUBLICATIONS/Sustainable_Hydrogen_Production_-_A_role_for_Fusion.pdf)

[26] Nuttall W J and Bakenne A T 2020 Will COVID-19 and cheap oil green our energy future? *Energy Policy Research Group* (Cambridge University) (https://eprg.group.cam.ac.uk/comment-will-covid-19-and-cheap-oil-green-our-energy-future-by-w-nuttall/)

IOP Publishing

Commercialising Fusion Energy
How small businesses are transforming big science
William J Nuttall, Satoshi Konishi, Shutaro Takeda and David Webbe-Wood

Chapter 12

Fusion energy and carbon management

Satoshi Konishi, Hoseok Nam and Shutaro Takeda

12.1 Fusion and carbon-based chemistry

Carbon neutral hydrogen production with fusion could be a promising option to replace fossil fuels. While fusion electricity can indeed produce hydrogen by electrolysis, however, it is not the most efficient process. Fusion energy can be recovered in the form of high-grade heat at a temperatures above that of light water fission reactors (325 °C, PWR) using various coolants such as liquid metal or helium gas. The use of advanced blanket concepts could add more advanced hydrogen technology such as Vapour Electrolysis or the Iodine-Sulfur cycle. However, the use of hydrogen as a fuel is still limited due to the lack of appropriate infrastructure, such as storage or delivery that need dedicated technology and administrative frameworks. Compounds of carbon such as hydrocarbons or alcohols as a substitute fuel have an advantage to deploy over the currently available supply chain, if the origin of the carbon involved is biomass then it may be regarded as neutral in terms of atmospheric emissions. Particularly with a high temperature blanket, fusion energy can be utilized for pyrolysis of biomass compounds, for example to produce gaseous or liquid fuels. The authors have reported a fusion reactor system design and proof of principle experiments based of this concept. In this chapter, the use of fusion energy conversion for the endothermic pyrolysis of biomass and its implications are discussed.

12.2 Pyrolysis and gasification of biomass

The processes of fusion and biomass fuel production rely on an endothermic reaction. Thermal heat is provided from the fusion blanket with high temperature media such as liquid lithium lead eutectic alloy (LiPb) with a nominal Li17%-Pb83% composition. Thc blankct conccpt such as DCLL (Dual Coolant Lithium Lead) applied in a Fuel Channel Insert (FCI) are made of ceramics. The ceramic thermally and electrically insulates the LiPb from the container made of Reduced Activation Ferritic/Martensitic Steel (RAFM). This approach has a temperature limit of

doi:10.1088/978-0-7503-2719-0ch12

around 500 °C and is subject to a magnetohydrodynamics (MHD) pressure drop. For even higher temperatures, Kyoto University has proposed the use of a SiC fibre reinforced SiC ceramic (SiCf/SiC) composite structure that enables the system to achieve over 900 °C of LiPb and hence He outlet temperature [1]. This structured SiCf/SiC composite has also successfully demonstrated heat exchange between 1000 °C LiPb and High Pressure He in a situation where helium is verified as the secondary medium technology to transfer high-grade heat. As shown in the figure 12.1, the LiPb blanket is connected to a heat exchanger, where coolants deliver the heat to the He fluid in heat exchangers and then supply heat to the biomass gasifier.

In order to consider the possible use of lower temperature heat output from the reactor, an ARIES-CS power plant design with a DCLL blanket has been discussed by Raffray *et al* [4]. Divertor and additional coolant He with the outlet pressure of 8 MPa from the ARIES-CS reactor goes through the steam boiler. This provides pre-heating for the steam gasification process. The gasifier performs its gasification of biomass at ambient or modest pressure, because it accepts wet solid biomass and biomass waste, and the pressure of the He heat source is approximately 3 MPa [5]. Biomass is anticipated to be crushed and separated from metallic or mineral impurities (by floating the feedstock in water) before supply.

Synthetic gas production from biomass is an endothermic reaction of 814 kJ (8.2 MJ kg^{-1}). It involves the conversion of the biomass to a H_2–CO mixture (a so-called 'syngas') at a temperature of over 700 °C. This can occur with a high reaction efficiency (equation (12.1)) where cellulose is considered as the representative of biomass. Lignin can be contained, but the endothermic reaction is essentially similar, i.e. biomass + steam makes syngas and external heat. The reaction equilibrium can proceed at a temperature above 700 °C, but higher temperatures are preferred in order to complete the gasification to a H_2–CO mixture so as to

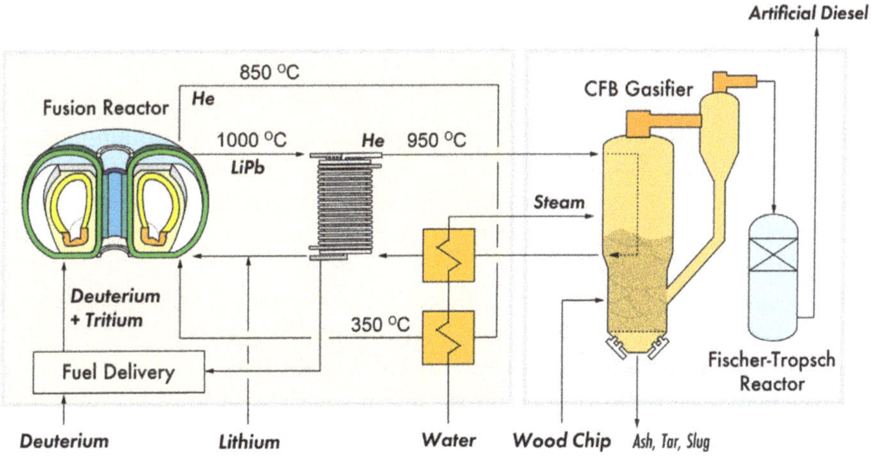

Figure 12.1. Schematics of biomass fuel production with fusion energy [2, 3].

minimize the production of distillate, tar and char. The syngas contains 24.2 MJ [6] of chemical energy considering the enthalpy of biomass is 16 MJ kg^{-1} [7].

$$n(C_6H_{10}O_5) + n/6(H_2O) \rightarrow nH_2 + nCO - 136n\ kJ\ mol^{-1} \quad (12.1)$$

This reaction implies large heat absorption by the endothermic reaction, and it can be expected that this will often be limited by the heat supply. It should be noted that, unlike in the case of a thermal cycle such as electricity generation by the Rankine cycle, this heat absorption process does not discard any waste heat, i.e. the thermal efficiency in this process is essentially close to unity. Considering the scenario where heat is to be generated from a fusion reaction, then use of heat for the endothermic reaction is approximately three times more effective than fusion electricity generation, or the related ideas of hydrogen production by electrolysis.

Syngas produced from the gasification process undergoes a gas cleaning process removing impurities for a higher conversion efficiency. The cleaned syngas contains a higher share of H_2 and CO compared to before gas cleaning. There are several options for the cleaned syngas to be converted into various energy carriers through chemical and technical processes. For example, when hydrogen is preferred, the shift reaction

$$H_2O + CO \rightarrow H_2 + CO_2 + 41\ kJ\ mol^{-1} \quad (12.2)$$

can readily be applied at lower temperatures in the train of the gas process.

The syngas produced by biomass gasification with the support of fusion energy may be converted into hydrocarbons of variable chain length via a Fischer–Tropsch (FT) reaction (equation (12.3)). The $-CH_2-$ made by the FT reaction is composed of hydrocarbons of different lengths, which are substitutions for petroleum gas, diesel, kerosene, jet fuel, etc. The synfuels generated by such processes can be immediately deployed in the existing infrastructure of the existing fuel supply chain, such as gas stations. Approximately 1 kg of waste biomass is ideally converted into 0.5 L of diesel fuel [6]. This process has not been broadly implemented in current petroleum-based chemical engineering, but it has been well-known for decades.

$$2H_2 + CO \rightarrow -CH_2 - + H_2O \quad (12.3)$$

Artificial liquid fuels produced via the Fischer–Tropsch reaction are a promising alternative for the long-term, because they include almost no contaminants (e.g., sulfur and aromatics, etc) compared to conventional gasoline and diesel, which results in low pollutant emissions [8]. Because biomass is regarded as 'carbon neutral', replacing petroleum fuel with biomass fuel results in saving fossil resource and reduction of CO_2 emissions.

In many cases, waste biomass from big cities or agricultural byproducts can easily be collected with minimal, or even negative, cost and effort.

Another way to use syngas is as a fuel for a solid oxide fuel cell (SOFC) for electricity generation or to produce hydrogen through water-gas shift reaction for the higher share of hydrogen in the syngas. Compared to other types of fuel cells, an SOFC is fuel-flexible. Many different types of gas such as natural gas, coal gas, biogas, hydrogen, etc can be used. Hence, either cleaned syngas or hydrogen can be a fuel for electricity generation by the SOFC.

It should be noted for fusion energy that still struggles to satisfy the Lawson criteria (see chapter 1), this biomass gasification process will make this problem approximately 10 times easier. Because of the endothermic nature of the reaction there is no waste heat, and the possible addition of external chemical heat from biomass, it is possible to enhance energy multiplication beyond the original fusion input by this application. This feature makes the realization of fusion considerably easier and brings it sooner in time. Also, small-scale fusion in the range of several hundred MW fusion power, or less, will have a possibility to achieve net positive output by this energy multiplication effect involving biomass gasification. Smaller-scale fusion devices are expected to have shorter lead-times and will require smaller financial investment in order to implement a viable fusion project.

12.3 Market for biofuel

Liquid fuel is expected to remain dominant in transportation energy consumption. The liquid fuel consumption of the global transportation sector is expected to account for 59% of demand in 2050 despite the improvements in on-road fuel economy. The OECD and non-OECD countries are estimated to consume approximately 50 and 80 quadrillion BTU of liquid fuels (i.e., residual fuel oil, diesel, jet fuel and motor gasoline) respectively in 2050 [9]. As indicated in figure 12.2, liquid fuels (except jet fuel) shows lower consumption in OECD countries in 2050 because of fuel economy and strict fuel efficiency standards. On the other hand, biofuel (i.e. distillate fuel oil and biodiesel) consumption in 2050 is expected to be higher than 2018 in non-OECD countries (up to 30 quadrillion BTU) despite the expected consumption drop in the OECD countries. Although natural gas and electricity are fast-growing energy carriers for both OECD and non-OECD countries, these together will still not outstrip the use of liquid fuel.

Retail diesel prices will be higher due to expected crude oil price rises, and its price range is between \$0.68 L^{-1} and \$1.85 L^{-1} based on the underlying crude oil price

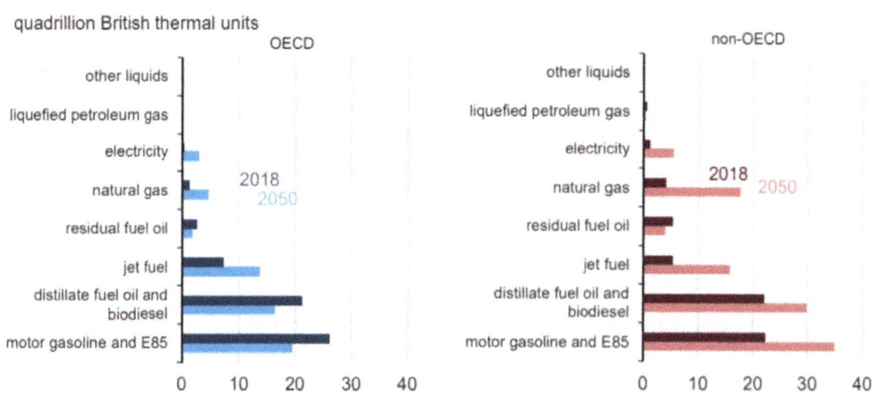

Figure 12.2. Transportation fuel consumption of OECD and non-OECD [9] (US EIA).

[10]. The percentage of biofuels blended into various types of liquid fuels in the US will be increased from 7.3% in 2019 to a peak of 9% in 2040 and then is expected to slightly decrease in 2050 [11]. In the case of a high oil price, the biofuel share is expected to rise to 13.5% in 2050. Compared to the electricity market, that will already be dominated by renewable and possibly nuclear power in the age when fusion will be introduced, the fuel market may have a larger demand for fusion energy.

12.4 Electricity generation by SOFC

Syngas itself, or hydrogen extracted from biomas syngas, may be used to generate electricity via the use of SOFC technology. The performance of the SOFC depends on electrochemical reactions under conditions of high temperature (> 600 °C). The SOFC reactions are described in equations (12.4)–(12.7). The syngas from the gas cleanup system enters the SOFC anode directly. As a SOFC operates at high temperature, both electricity and heat are generated for energy use. The by-product heat from an SOFC is of high quality.

$$\text{Anode} \qquad H_2 + O_2 - \rightarrow H_2O + 2e- \tag{12.4}$$

$$CO + O_2 - \rightarrow CO_2 + 2e- \tag{12.5}$$

$$CH_4 + 4O_2 - \rightarrow CO_2 + 2H_2O + 8e- \tag{12.6}$$

$$\text{Cathode} \qquad O_2 + 2e - \rightarrow O_2- \tag{12.7}$$

Oxygen is provided to the porous cathode where it turns to oxygen ions. Then, the oxygen ions move to the anode through the solid electrolyte. At the anode, oxygen ions can combine with hydrogen and produce water along with carbon dioxide. For the SOFC component, hard and rigid materials are necessary because of the high operating temperature. SOFC efficiency is broad between 40% and 90% according to various factors including: the system scale, operation, materials, and design [12].

Electricity generated from biomass synfuel can be used for mid- to large-scale generation, and also at extremely small-scale such as for residential use. In both cases, SOFC electricity is expected to respond the changing demand with good efficiency due to the co-generation that utilizes heat. This is even more preferable for the future grid which can be expected to have less flexibility because of the larger fraction of renewable and nuclear, both of which do not respond to the system demand [13]. Also, for smaller electricity consumers with limited grid access, such as in developing countries, the provision of biomass derived synfuel could eventually contribute to electricity supply.

12.5 Effect of subsidy

We provide an economic analysis of the comparison between electricity generation from SOFC and artificial diesel production using fusion energy (figure 12.3). Diesel and electricity prices are estimated at \$1.08 L^{-1} and \$110 MWh^{-1}, respectively [10]. Under the assumption that the fusion energy cost is the same for syngas production,

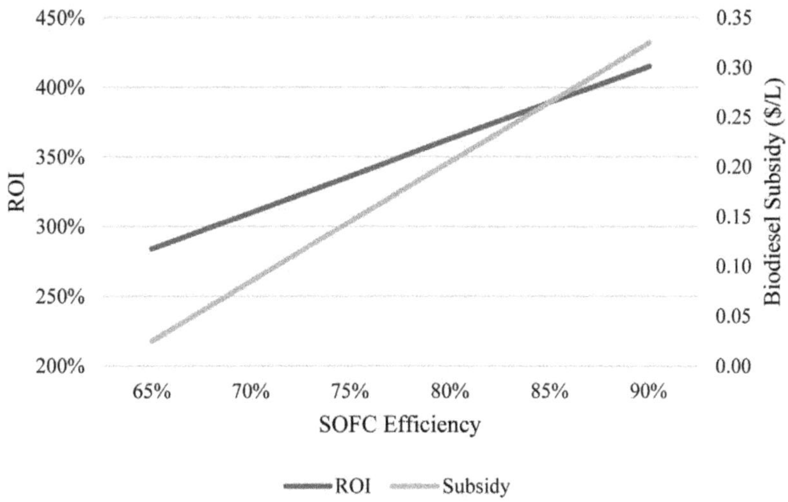

Figure 12.3. Diesel subsidy pricing according to SOFC efficiency [3].

different costs for a diesel production facility and a SOFC facility are compared. The return on investment (ROI) for the SOFC proposition is described on the left side of *y*-axis according to the SOFC efficiency (*x*-axis). The right side of *y*-axis indicates how much subsidy for the diesel is required, so as to adjust to the same ROI as the SOFC noting the efficiency changes. The result demonstrates that the necessary subsidy ranges from \$0.03 L^{-1} to \$0.33 L^{-1} according to the SOFC efficiency changes between 65% and 90%.

The subsidy for diesel production is varied by the policies, energy usage, and feedstock availability of each country. This study observes that the US optimal diesel subsidy is \$0.92 $gallon^{-1}$ (\$0.2 L^{-1}) [14]. Diesel subsidies in the EU shows between €0.15 L^{-1} and €0.4 L^{-1} [15]. Subsidies in Canada and Switzerland are seen to be \$0.2 L^{-1} and \$1 L^{-1}, respectively [16]. Also, Japan and Australia implement \$0.28 L^{-1} and \$0.31 L^{-1}, subsidies respectively [17]. The subsidy in Indonesia is \$0.37 L^{-1} after an increase in 2015 [18].

12.6 Fusion charcoal production

One of the merits of the fusion energy is that it is a low-carbon energy source, something which is essential for effective climate action. To maximize the merit, charcoal production from biomass using fusion energy can be a carbon-negative industry. That is because the charcoal produced can be considered as stable and fixed carbon not emitted to the atmosphere. Low temperature heat is required to produce charcoal, even a water-cooled blanket, which provides a relatively low outlet temperature of about 280 °C may usefully be applied. By capturing solid carbon in this way, it has the possibility to be traded via an emission trading scheme (ETS) as part of a future geared to CO_2 reduction. Currently in the European Union the ETS is implemented at city, regional and national levels. By the time of fusion commercialization in the year 2050, the ETS concept is likely to be expanded across

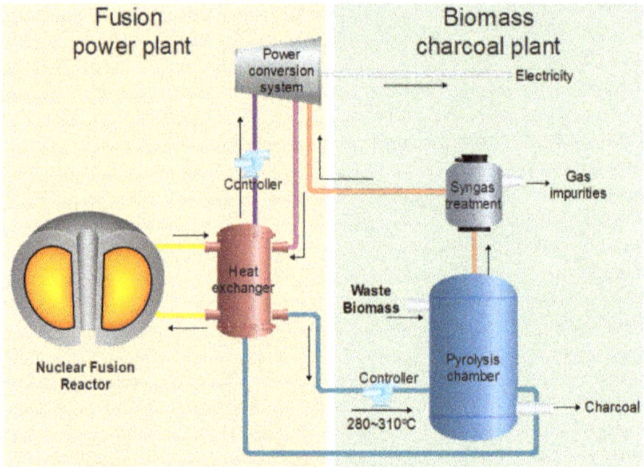

Figure 12.4. Schematics for of charcoal production with fusion energy (Nam *et al* [20] with permission of Springer).

the world. Through the ETS and similar mechanisms, profit can be made from emission credit trading besides the sale of electricity and fuels.

The chemical formula for the conversion of biomass-to-charcoal process is described [19]. The biomass is assumed to be woody and contains large proportion of lignin (equation 12.8).

$$(CH_{1.6}O_{0.6})n = nC + 0.2nH_2 + 0.6n\ H_2O - 451n\ kJ \tag{12.8}$$

Charcoal as a solid carbon can be separated through the endothermic reaction with 451 kJ mol^{-1} of heat. The stoichiometric calculation indicates that the biomass weight of $CH_{1.6}O_{0.6}$ is 23.2 g mol^{-1}, which the amount of endothermic heat requires 19.4 MJ kg^{-1}.

Regarding the charcoalization process, a heat transfer agent (helium) from the heat exchange delivers low temperature heat between 280 °C and 310 °C. If a higher temperature from the varied blanket designs is provided, then the charcoal can be sintered. As shown in figure 12.4, helium fluid from the heat exchanger circulates to the biomass chamber for the slow pyrolysis. After the pyrolysis process, the gaseous mixture from the pyrolysis chamber goes through the syngas treatment process for the removal of impurities. Then, the cleaned gas enters the power conversion system for electricity, which is used for internal power for the operation of the system.

12.7 Market in carbon credits

As of 2019, a total of 57 carbon pricing initiatives have been implemented or been scheduled for implementation. Among them, the 28 EU member state ETS system is implemented in regional, national and subnational jurisdictions, while 29 carbon taxes are also applied on a national level (both traded permits and taxes are examples of carbon pricing). As shown in figure 12.5, the ETS is based on the volume of greenhouse gas (GHG) reduction, meaning that it sets mandatory limits on the CO_2 equivalent emission. Then, emission quotas or allowances (tradable emission permits) are distributed to the emitters. If participants in the market have

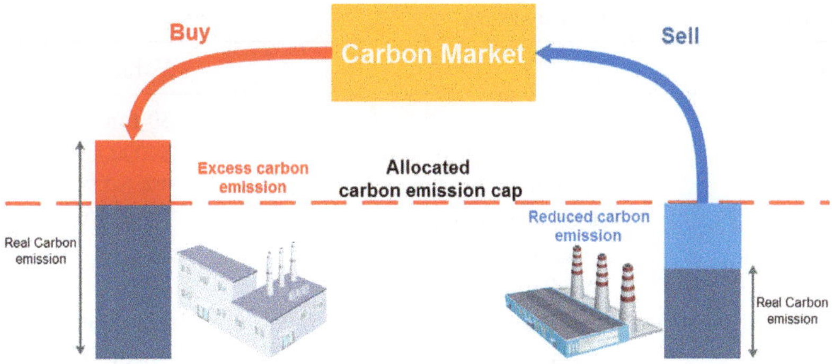

Figure 12.5. Illustration of emission trading scheme.

insufficient allowances, they can purchase in the market to compensate for their excessive emissions. In addition, allowances can be sold to the excessive emitters as rewards for additional reduction efforts. This market enables participants to lower the economic cost of reducing overall emissions. In principle, emission reductions will be made most strongly by those best placed to do so.

12.8 Economic analyses of CCS

A previous study on charcoal production cost for carbon sequestration and the breakeven cost according to the fusion cost have been performed [20]. To discover the economic feasibility of the fusion energy and charcoal production, economic analysis of return on investment and payback period are calculated. Major parameters (table 12.1) are imported from Nam *et al* [20].

The net present value (NPV) is the cumulative value (revenues—expenses) adjusted to the present time. The discount rate (i) that renders the net present value (NPV) of all cash flows to be zero. As it takes the time-value of expenses and revenues over the life period of the project into consideration, a practical evaluation of the returns can be provided shown in (equation 12.9). Annual net (after-tax) profit is determined that income is deducted by expenses considering tax and depreciation (equation 12.9).

$$\mathrm{NPV} = \sum_{t=0}^{n} \frac{\mathrm{Annual\ Net(After - tax)Profit}}{(1 + i)^t} \tag{12.9}$$

Discounted cash flow return on investment (DCF ROI) is a well-established method to consider the changing value of money over time in order to evaluate/compare the efficiency of investments (equation 12.10). It directly measures the amount of return (or benefit) on an investment divided by the cost of the investment for fusion energy and charcoal production plant (Fubio.TCI). The result is demonstrated as a percentage, or a ratio.

$$\mathrm{DCF\ ROI} = \frac{\mathrm{NPV}}{\mathrm{Fubio.TCI}} \times 100 \tag{12.10}$$

Table 12.1. Parameter for economic analysis [20].

Parameter		Value
Waste biomass		\$0 ton^{-1}
Fusion power plant	Total Capital Investment (TCI)	\$4000 kW_e^{-1}
	Operating & Maintenance (O&M)	3% of TCI
	Electrical output	1000 MW_e
	Electrical efficiency	33%
	Availability	75%
Biomass charcoal plant	Total capital investment	\$55 million
	Operating & maintenance	9% of TCI
	Capacity	470 ton day^{-1}
Common parameter for fusion-biomass hybrid system	Operating years	30 years
	Discount rate	7%
	Income tax	30%
	Depreciation	MACRS using a 15-year recovery period

The discounted payback period (DPBP) provides the number of years it takes to break even, considering the initial investment. The DPBP equation is not complex, as it can be described that the discounted cash flow discounts each inflow until NPV equals zero (equation (12.11)).

$$\text{DPBP} = t \text{ when NPV} = 0 \tag{12.11}$$

where t is the number of years since the start of operation.

In terms of the operation and maintenance (O&M) cost, it is reported that O&M cost can be up to 13% of fusion TCI due to the increase in blanket component replacement [21] (figure 12.6). A literature survey reveals that carbon sequestration costs from char production range \$37–92 tCO_2^{-1} [22, 23]. Hence, the fusion energy and charcoal production method has the potential for competitive deployment if the fusion technology cost is low enough to compete with the conventional methods for char production (figure 12.7).

It is found that an emission credit lower than \$50 tCO_2^{-1} is not sufficient to make the charcoal production with fusion energy attractive, meaning that a net deficit can be expected after 30 years of operation. In case of \$80 tCO_2^{-1} of the emission credit price, the DCF ROI is 11% at 3% of the fusion O&M cost, while the DCF ROI is 12% at 9% of the fusion O&M cost under 110 \$ CO_2^{-1} of the emission credit price. If the emission credit price is higher than \$140 tCO_2^{-1}, then the DCF ROI shows all positive values and it is only matter of how much profit can be acquired compared with the initial investment after the operation termination.

Figure 12.8 shows the DPBP as function of emission credit prices (x-axis) and variable combination of fusion costs (line type). Sensitivity result shows that not all, but most, fusion cost combinations enable the recovery of the necessary investment

$$\text{DPBP} = \text{t when NPV} = 0 \quad (1.3.4)$$

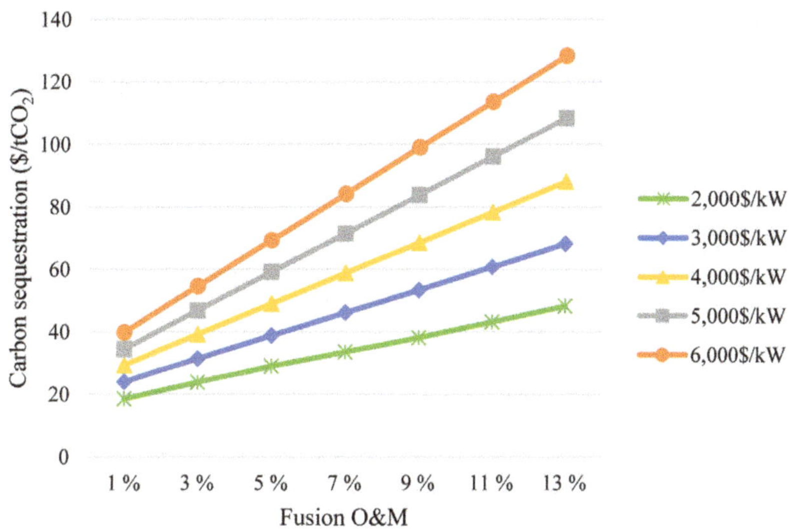

Figure 12.6. Carbon sequestration cost by charcoal production based on the fusion cost [20].

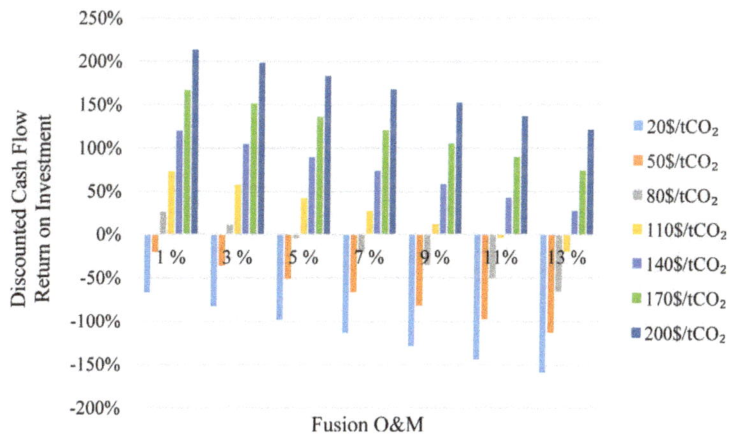

Figure 12.7. Discounted cash flow returns on investment according to O&M change at default \$4000 kW_e^{-1}.

cost within approximately 15 years out of a total of 30 years of operation. The DPBP reacts more sensitively to the emission credit price when the fusion O&M is high. For example, a \$2000 kW^{-1} at 13% O&M shows a lower DPBP than \$3000 kW^{-1} at 1% O&M at \$140 tCO_2^{-1} of emission credit price. Also \$3000 kW^{-1} at 13% O&M shows faster investment recovery than \$5000 kW^{-1} at 1% O&M at \$140 tCO_2^{-1} of emission credit price. Therefore, when the emission credit price is low, low fusion TCI with high O&M cost shows a better investment recovery than high fusion TCI with low O&M cost, and vice versa. The best practice for the lowest DPBP is to lower both fusion TCI and O&M cost.

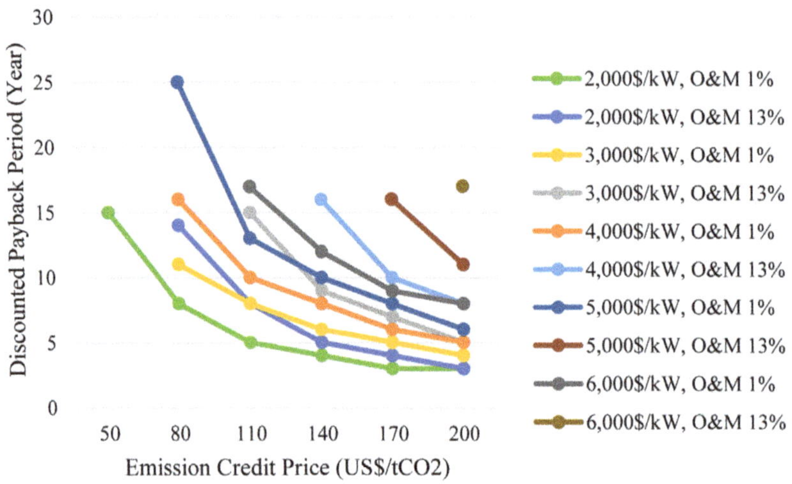

Figure 12.8. Discounted payback period according to the emission credit price and fusion cost. Note: DPBP over 30 years do not appear in the figure.

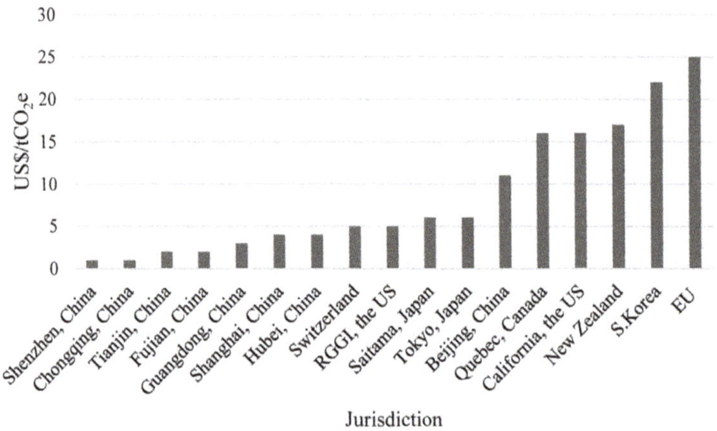

Figure 12.9. Nominal prices in the ETS 2019. Note: The unavailable date for the ETS price is not shown in the figure.

12.9 Carbon credit trading

As described in figure 12.9, carbon prices in the carbon market vary from less than \$1 tCO_2e^{-1} to \$25 tCO_2e^{-1}. The EU price shows the highest price, \$25 tCO_2e^{-1} followed by \$22 tCO_2e^{-1} in South Korea. The EU carbon market has the largest source of revenues because of its size. In terms of energy sector in the EU, a massive surplus of allowances discourages the incentive for low-carbon investment [24]. China operates 8 ETSs in a city and regional level. Although the prices are not high enough (\$1 tCO_2e^{-1} to \$11 tCO_2e^{-1}) to significantly affect investors for firm investment decisions, China has the second-largest carbon market next to the EU. Chinese national and regional ETSs encourage the improvement and optimization of production processes in the industrial sector by raising awareness of GHG emissions. The consensus across many countries is that the main objective of policy is to achieve cost-effective emissions mitigation. Beyond that low-carbon intention, each country has deployed carbon pricing to differing extents. For instance, China [25], Korea [26], and Quebec [27], etc, support low-carbon technology and innovation, while Beijing also particularly wants to improve air quality [28].

The jurisdictions where ETS is established show a tendency to gradually increase the emission credit price linking and/or to cooperate with others [29]. In addition, trans-national banks have announced that they will intensify the carbon pricing in the future as follows [30]; Asian Development Bank schedules to increase the carbon price annually by 2% based on the carbon price of \$36.3 tCO_2^{-1} in 2016. The European Investment Bank has a plan to raise the carbon prices from \$47 tCO_2^{-1} to \$150 tCO_2^{-1} between 2018 and 2050 according to its climate action strategy. The World Bank has disclosed low and high cases of carbon price \$50 tCO_2^{-1} and \$100 tCO_2^{-1}, respectively, for 2030. The International Finance Corporation has suggested carbon pricing rising to a possible \$80 tCO_2^{-1} by 2050. The emission credit price was not directly discussed by those trans-national banks, but these announcements emphasize the significance of carbon pricing to long-term global finance. The intention is to enlarge the carbon pricing to a far broader range of sectors. This can be achieved while ensuring that the emission credit price is likely to increase.

12.10 Biomass feedstock and supply chain

12.10.1 Residential garbage and its collection

The World Bank has released two reports on the global solid waste (figure 12.10). The result on the global waste generation projection indicates that the world is expected to generate 2.22 billion tonnes of waste annually by 2025 [31]. By 2030, it is estimated that 2.59 billion tonnes of waste will be generated, and reach to 3.4 billion tonnes in 2050 [32].

Comparing data from the two reports, organic waste (food, etc) decreases from 46% in 2010 to 44% in 2016 (figure 12.11). Despite the fact that the increase in both organic waste generation in high-income countries, a decrease in lower-middle- and low-income countries leads to lower organic waste generation. However, an increase in plastic waste generation in high-income countries makes the global plastic generation increase, although lower-middle- and low-income countries reduced plastic waste. The sum of wood, rubber and leather, and other is 18%, while others in 2010 account for 18%, which is considered to include timber, and rubber and leather. In this sense, the shares in 2010 and 2016 do not show differences.

Approximately 33% of waste is disposed of in open dumps, which is highly subject to producing greenhouse gas impacting climate change (figure 12.12). When looking inside, low and lower-middle-income countries show a high share of 93% and 66%, respectively. This is a big difference compared to the high-income countries which

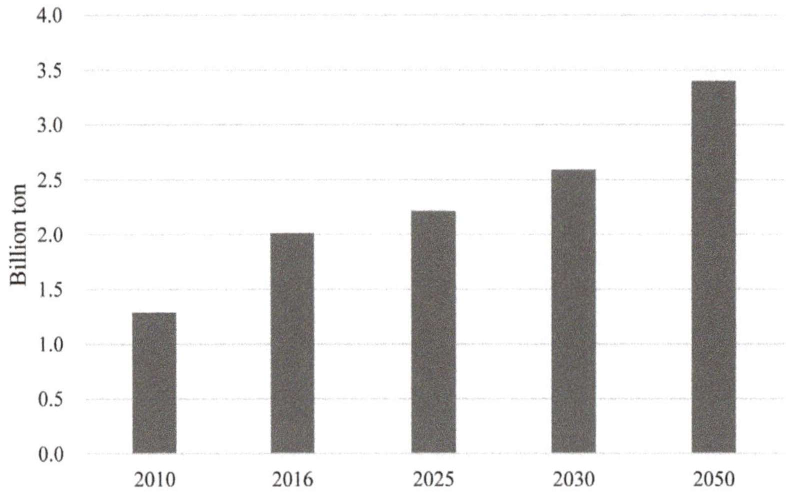

Figure 12.10. Solid waste generation and its projection [31, 32].

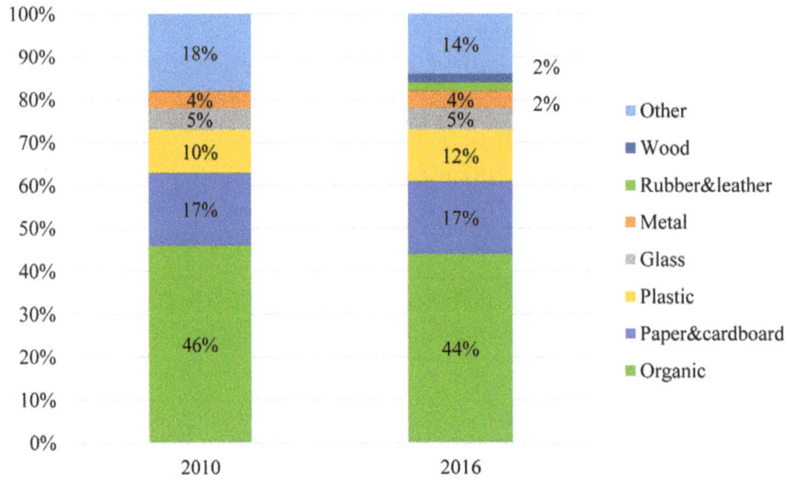

Figure 12.11. Composition of the global solid waste in 2010 and 2016 [31, 32].

shows only 2% of open dumps. Regarding landfills, 25% is an unspecified landfill, which as well has a high risk of contaminating water by leaking and produce greenhouse gas by decomposition. Recycling and composting for material recovery accounts for 18.5%, and 11% is managed through modern incineration.

About 55% of waste treatment is managed in the anti-environment method. Considering 2.01 billion tonnes of waste was generated in 2016, about 1.11 billion tonnes of waste was treated in unfavourable ways for the environment. In this context, 1.87 billion tonnes of waste in 2050 will be managed in open dumps and uncontrolled landfill, conveying a negative impact on the environment.

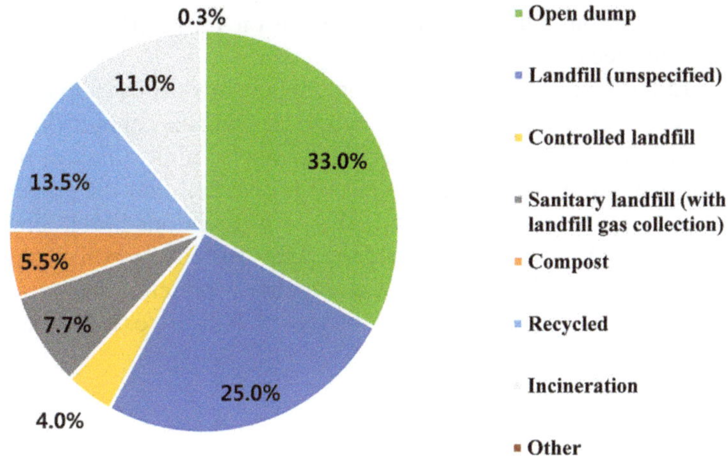

Figure 12.12. World waste treatment and disposal [32].

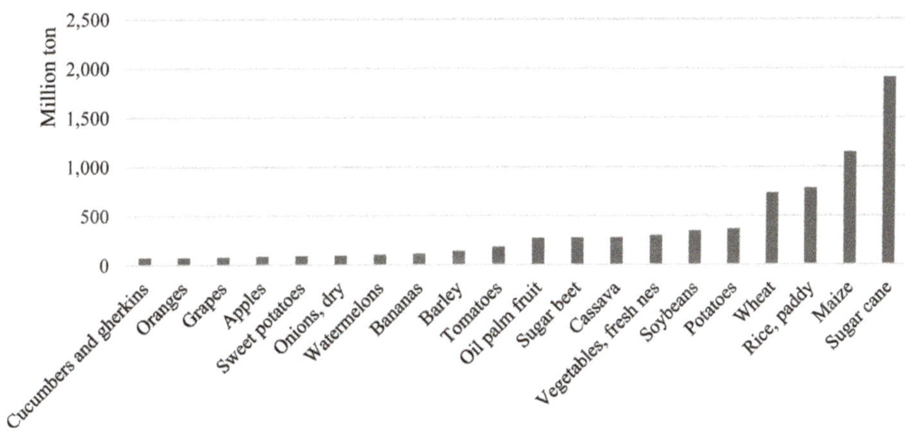

Figure 12.13. World top 20 crop production in 2018 [33].

12.10.2 Agricultural by-products and their characteristics

The significance of crop production is related to harvested areas and yields per area (figure 12.13). The increase in crop production has been driven by a significant increase in yields per areas with crop intensification. However, trends are not equally shown in all regions. The highest crop production is sugar cane (1907 million tons), mainly from Latin and Central America and Asian countries. Wheat and rice productions are 734 million tons and 782 million ton, respectively, both of which are mainly from Asia and Northern Africa. Expansion of harvested land has led to the production growth of maize (1147 million tons) in Latin America and sub-Saharan Africa.

Since the world crop production shows high quantities, the residues quantities after harvesting and processing cannot be overlooked. The residues may be able to be used as fertilizer, but might cause CO_2 emission returning to the atmosphere when it spoils and decays (table 12.2).

According to the literature review, carbon content in various crop residue can be observed. Depending on the regions and types, the carbon content differs slightly. The average carbon content is 47.3% which accounts for almost half of the chemical composition in the crop residues (table 12.3). It is reported that the polishing rate of rice and barely is about 75%, indicating that 25% is abandoned [41]. Since different crops generate varied amounts of residues with varied carbon content, better utilization of crop residues has been studied. Biochar production using crop residues can be applied as soil improvement for plant growth [42]. Also, it can be applied for water and air purification [43, 44].

Table 12.2. Ultimate analysis of different crop residues.

Crop residue	C	H	O	Ref.
Almond shell	51.4	6.1	41.6	[34]
Apricot tree pruning	51.4	6.3	41.2	[35]
Bagasse	43.8	5.8	47.1	[36]
Barley straw	46.8	5.5	41.9	[35]
Coconut coir	47.6	5.7	45.6	[36]
Coconut shell	50.2	5.7	43.4	[36]
Coir pith	44.0	4.7	43.4	[36]
Corn cobs	47.6	5.0	44.6	[36]
Corn stalks	45.5	6.2	41.1	[35]
Cotton gin waste	42.7	6.0	49.5	[36]
Cotton stalks	41.2	5.0	34.0	[35]
Durian shell	60.3	8.5	28.1	[37]
Ground nut shell	48.3	5.7	39.4	[36]
Millet husk	42.7	6.0	33.0	[36]
Oats straw	46.0	5.9	43.5	[35]
Olive tree pruning	49.9	6.0	43.4	[35]
Palm shell	50.1	6.9	41.2	[38]
Peach tree pruning	53.0	5.9	39.1	[35]
Peanut shell	46.8	6.6	37.6	[39]
Plum stones	46.4	5.5	48.0	[40]
Rice husk	38.9	5.1	32.0	[36]
Rice straw	41.8	4.6	36.6	[35]
Subabul wood	48.2	0.0	45.1	[36]
Sugar beet leaves	44.5	5.9	42.8	[35]
Sunflower straw	52.9	6.8	35.9	[35]
Vineyard pruning	47.6	5.6	41.1	[35]
Wheat straw	47.5	0.1	35.8	[36]
Average	47.3	5.4	40.9	

Table 12.3. Biomass feedstock prices.

Biomass feedstock	Price ($ ton^{-1})	Point in supply chain	Ref.
Bio-fraction of MSW	−53 (−60 to −31)	Gate fee	[45]
Bio-fraction of commercial and industrial waste	−53 (−60 to −13)	Gate fee	
Animal manure	0 (0 to 44)	Farm gate	
Sewage sludge	0 (−53 to 0)	Water treatment works	
Palm oil mill effluent	0	Mill	
Empty palm fruit bunches	4 (3 to 5)	Mill	
Bagasse	11 (4 to 44)	Sugar mill	
Bark, branches, leaves	51 (44 to 57)	Delivered to buyer	
Corn stover	51	Farm gate	
Solid derived fuel	55	Roadside	
Short rotation coppice	65	Farm gate	
Miscanthus	69	Farm gate	
Grape marcs	70	Dried and delivered	
Straw	82 (62 to 98)	Farm gate	
Nut shells	87 (64 to 111)	Factory gate	
Saw dust and cutter shavings	87	Mill	
Forest residues	23 (15 to 30)	N/A	[46]
Wood waste	30 (10 to 50)	N/A	
Agricultural residues	35 (20 to 50)	N/A	

Although these biomass resources like wood chips and corn, etc have varied costs, economic values can be higher after proper chemical and physical processing. Even more, waste biomass of residues and municipal solid waste, which does not have any value, can become a valuable asset to satisfy energy demand. Waste biomass shows a broad range of prices, as described in tables 12.1 and 12.2. To utilize the waste biomass at zero or low cost, it is necessary to take a pre-measurement by a long-term contract, etc, because high demand omcreases prices. It should be noted that the feedstock price is varied depending on the cost structure. For example, forest residues include harvesting and stumpage cost, while it is mentioned that wood waste can be zero because disposal cost is high. Wastes from the city and farm are negative prices, while others, which can be used for other commercial purposes besides energy production, show quite high prices.

12.11 Summary

In this chapter, the authors proposed artificial fuel production from biomass feedstock with fusion as a promising alternative to replace fossil fuels. Through this process, fusion can provide much-needed carbon neutral fuel to the market. Among possible options, the gasification process combined with the Fischer–Tropsch process may be most promising for the clean and efficient use of biomass resources. The key requirements for the gasification process are high temperature and a large amount of heat. Fusion reactors, when designed for the purpose, can provide both. With this biomass process, the breakeven conditions for nuclear fusion reactors can be eased. The authors calculate that net positive energy gain could be achieved with $Q_{\text{fus}} \sim 5$. This would speed up the realization of commercial fusion.

Furthermore, as a more advanced concept, a charcoal production process with heat from fusion reactors was also discussed. Through this option, the revenue will entirely be generated from the carbon market. As the world races toward decarbonization, this novel option may be able to achieve economic advantages over other conventional uses of fusion energy. The author provided a detailed analysis of the economy and the supply chain of this concept to make a case for this innovative idea.

References

[1] Ibano K, Utoh H, Tobita K, Yamamoto Y and Konishi S 2011 Design studies of innovatively small fusion reactor based on biomass-fusion hybrid concept: GNOME *Fusion Eng. Des.* **86** 2779–82

[2] Takeda S 2019 *Biomass Energy Carbon Capture and Sequestration with Nuclear Fusion: A Promising Option for Future Energy Sustainability?* (Cambridge, MA: Harvard University)

[3] Nam H, Kasada R and Konishi S 2018 Economic analysis between diesel and SOFC electricity via fusion-biomass hybrid model *J. Fusion Energy* **37** 333–45

[4] Raffray A, El-Guebally L, Malang S, Wang X, Bromberg L, Ihli T, Merrill B and Waganer LARIES-CS team 2008 Engineering design and analysis of the ARIES-CS power plant *Fusion Sci. Technol.* **54** 725–64

[5] Larson E D, Jin H and Celik F E 2009 Large-scale gasification-based coproduction of fuels and electricity from switchgrass *Biofuel. Bioprod. Biorefin.* **3** 174–94
[6] Konishi S, Ichinose M, Ibano K and Yamamoto Y 2010 Fusion-biomass hybrid conept and its implication in fusion development *23rd IAEA Fusion Energy Conf.*
[7] European Commission 2003 *Refuse Derived Fuel, Current Practice and Perspectives* (Brussels: European Commission-Directorate General Environment)
[8] Tijmense M J, Faaij A P, Hamelinck C N and van Hardevel M R 2002 Exploration of the possibilities for production fo Fischer Tropsch liquids and power via biomass gasification *Biomass Bioenerg.* **23** 129–52
[9] EIA 2019 *International Energy Outlook 2019 with Projections to 2050* (Washington, DC: U. S. Energy Information Administration)
[10] EIA 2018 *Annual Energy Outlook 2018 with Projections to 2050* (Washington, DC: U.S. Energy Information Administration)
[11] EIA 2020 *Annual Energy Outlook 2020 with Projections to 2050* (Washington, DC: U.S. Energy Information Administration)
[12] Irshad M, Siraj K, Raza R, Ali A, Tiwari P, Zhu B, Rafique A, Ali A, Ullah M K and Usman A 2016 A brief description of high temperature solid oxide fuel cell's operation, materials, design, fabrication technologies and performance *Appl. Sci.* **6** 1–23
[13] Takeda S, Sakurai S, Yamamoto Y, Kasada R and Konishi S 2016 Limitation of fusion power plant installation on future power grids under the effect of renewable and nuclear power sources *Fusion Eng. Des.* **109** 1754–8
[14] Wu H, Colson G, Escalante C and Wetzstein M 2012 An optimal U.S. biodiesel fuel subsidy *Energy Policy* **48** 601–10
[15] Global Subsidies Initiative 2013 *Addendum to Biofuels—At What Cost? A Review of Costs and Benefits of EU* (Winnipeg: The International Institute for Sustainable Development)
[16] Wolf M 2007 Biofuels: a tale of special interests and subsidies *Financial Times* (https://ft.com/content/40a71f96-8702-11dc-a3ff-0000779fd2ac)
[17] G S Initiative 2007 *Biofuels—At What Cost?: Government Support for Ethanol and Biodiesel in Selected OECD Countries* (Winnipeg: The International Institute for Sustainable Development)
[18] USDA 2016 *Indonesia Biofuels Annual 2016* (Washington, DC: US Department of Agriculture Foreign Agricultural Service)
[19] Konishi S, Takeda S and Nam H 2018 Future possibility of carbon sequestration by biomass fusion hybrid systems *27th IAEA Fusion Energy Conf.* (Gujara, India: Mahatma Mandir)
[20] Nam H, Mukai K, Yagi J and Konishi S 2020 Economic analysis of carbon removal strategy by fusion–biomass hybrid system in the future market mechanism *J. Fusion Energy* (Accepted)
[21] Gi K, Sano F, Akimoto K, Tobita K and Hiwatari R 2018 Contribution of fusion energy to low-carbon development under the Paris Agreement and accompanying uncertainties: FIP/P3-2 *27th IAEA Fusion Energy Conf.-IAEA CN-258 (Gandhinagar, India)*
[22] Timmons D, Lema-Driscoll A and Uddin G 2017 *The economics of biochar carbon sequestration in Massachusetts* (Boston, MA: University of Massachusetts Boston/UMass Clean Energy Extension)
[23] Lehmann J 2007 A handful of carbon *Nature* **447** 143–4
[24] World Bank 2016 *State and Trends of Carbon Pricing* (Washington, DC: World Bank Group)

[25] NDRC 2017 National carbon emission trading system started (in Chinese) *National Development and Reform Commission* (http://ndrc.gov.cn/xwzx/xwfb/201712/t20171219_871024.html)
[26] Greenhouse Gas Inventory and Research Center 2017 Current status of the Korean ETS *Greenhouse Gas Inventory and Research Center* (https://thepmr.org/system/files/documents/PMR_Country%20presentation_20%20Mar%202017_EN_final_revised.pdf)
[27] Quebec 2018 The Quebec cap and trade system for greenhous gas emission allowances (http://mddelcc.gouv.qc.ca/changements/carbone/Systeme-plafonnement-droits-GES-en.htm)
[28] BMPC 2014 Report on the status of the city's carbon emissions trading pilot project (in Chinse) *The Standing Committee of Beijing Municipal People's Congress* (http://bjrd.gov.cn/zdgz/zyfb/bg/201403/t20140321_129641.html)
[29] World Bank 2018 *State and Trends of Carbon Pricing 2018* (Washington, DC: World Bank Group)
[30] Carbon Pricing Leadership Coalition 2018 *Carbon Pricing Leadership Coalition Leadership Report 2017–2018* (Washington, DC)
[31] Hoornweg D and Bhada-Tata P 2012 *What a Waste: A Global Review of Solid Waste Management. Urban Development Series: Knowledge Papers No. 15* (Washington, DC: World Bank)
[32] Kaza S, Yao L C, Bhada-Tata P and Van Woerden F 2018 *What a Waste 2.0: A Global Snapshot of Solid Waste Management to 2050. Urban Development* (Washington, DC: World Bank)
[33] FAOSTAT 2020 *Data: crops. Food and Agriculture Organization of the United Nations* (http://fao.org/faostat/en/#data/QC)
[34] Plaza M, Pevida C, Martin C, Fermoso J, Pis J and Rubiera F 2010 Developing almond shell-derived activated carbons as CO2 adsorbents *Sep. Purif. Technol.* **71** 102–6
[35] Encinar J, Gonzalez J and Gonzalez J 2002 Steam gasification of *Cynara cardunculus* L.: influence of variables *Fuel Process. Technol.* **75** 27–43
[36] Kirubakaran V, Sivaramakrishnan V, Nalini R, Sekar T, Premalatha M and Subramanian P 2009 A review on gasification of biomass *Renew. Sustain. Energy Rev.* **13** 179–86
[37] Chandra T, Mirna M, Sudaryanto Y and Ismadji S 2009 Activated carbon from durian shell: preparation and characterization *J. Taiwan Inst. Chem. Eng.* **40** 457–62
[38] Daud W, Ali W and Sulaiman M 2000 The effects of carbonization temperature on pore development in palm-shell-based activated carbon *Carbon* **38** 1925–32
[39] Wang S, Nam H and Nam H 2020 Preparation of activated carbon from peanut shell with KOH activation and its application for H2S adsorption in confined space *J. Environ. Chem. Eng.* **8** 103683
[40] Nowicki P, Wachowska H and Pietrzak R 2010 Active carbons prepared by chemical activation of plum stones and their application in removal of NO_2 *J. Hazard. Mater.* **181** 1088–94
[41] Nam H and Konishi S 2019 Potentiality of biomass-nuclear hybrid system deployment scenario: techno-economic feasibility perspective in South Korea *Energy* **175** 1038–54
[42] Maraseni T N 2010 Biochar: maximising the benefits *Int. J. Envrion. Stud.* 319–27
[43] Saucier C, Adebayo M A, Lima E C, Cataluna R, Thue P S, Prola L D, Puchana-Rosero M, Machado F M, Pavan F A and Dotto G 2015 Microwave-assisted activated carbon from cocoa shell as adsorbent for removal of sodium diclofenac and nimesulide from aqueous effluents *J. Hazard. Mater.* **289** 18–27

[44] Wang S, Nam H and Nam H 2019 Utilization of cocoa activated carbon for trimethylamine and hydrogen sulfide gas removals in a confined space and its techno-economic analysis and life-cycle analysis *Environ. Prog. Sustain. Energy* e13241

[45] Arup URS Consortium 2014 Advanced biofuel feedstocks—an assessment of sustainability (https://assets.publishing.service.gov.uk/government/uploads/system/uploads/attachment_-data/file/277436/feedstock-sustainability.pdf)

[46] IRENA 2012 *Renewable Energy Technologies: Cost Analysis Series: Biomass for Power Generation* (Abu Dhabi: International Renewable Energy Agency)

Chapter 13

Conclusions

William J Nuttall, Satoshi Konishi, Shutaro Takeda and David Webbe-Wood

In this book, we have seen that the field of fusion science and technology is in the midst of profound change. This change has several dimensions, including: the increased involvement of the private sector in fusion development; that large publicly-funded projects are actively being prepared in anticipation of commercial technology demand; and that the energy industry has started to consider a role for fusion as part of their own transition to a low-carbon economy.

Fusion has tended to see technology requirements as presenting challenges in the decades' long journey from fusion science to fusion engineering. However, technology is increasingly finding a role as an enabler, whether that is the materials science of high-temperature superconductivity facilitating the development of high field tokamak devices or the use of computer vision and rapid digital control for the precise positioning of targets in inertial confinement fusion systems. Another shift underway is visible in investors' attitudes, be they wealthy individuals concerned for the climate emergency or financial institutions conscious of the fiduciary duties to mitigate risks related to our energy future. Finally, we are seeing a shift in individual enthusiasm. Increasingly young researchers and technologists are attracted to the sheer technical difficulty of fusion. Once the realization dawns that fusion may not be as difficult as it was once perceived to be, then their motivations start to shift from bold ambition to quiet confidence.

As one enters the development labs emphasized in the pages of this book, one does not encounter researchers working on decades-long projects destined for success in the distant future; rather, one has a sense of immediacy concerning the challenges and the opportunity.

We trust that this book has managed to make the case that the third decade of the 21st century has the prospect of seeing genuine progress in fusion commercialization. It seems probable that any such progress will be based upon fundamental physics that has been known for more than 50 years but was impossible to engineer until recently. Reading through this book, readers may find key enabling

doi:10.1088/978-0-7503-2719-0ch13

technological innovations that could be relevant to fusion commercialization including, but not restricted to:

- Practical high-temperature superconductors;
- For experiments: high-speed data processing and responsive instrumentation;
- For theory: affordable high-powered computing for modelling;
- Computer-aided design and manufacture;
- Advanced materials for hostile environments.

It would be wrong, however, to conclude that the emergence of commercial fusion enterprises has simply been a consequence of new materials and technologies. We suggest that the timing is not accidental and hence we posit that the following historical trends have been important.

- The early 1990s saw the end of the Cold War and the winding down of much of the military-industrial complex. This combined with a wave of economic liberalization that saw the privatization of many state-owned industries and the closure of many national laboratories, especially in the energy sector.
- The late 1990s were known for the emergence of a range of internet companies in the 'dot-com' boom. These companies absorbed investment capital, but much was lost as the boom became a bubble, and then it burst. Russia, once a superpower, was having a difficult time.
- The period 2000–2010 saw the 9/11 terrorist attacks on the United States and the subsequent 'War on Terror.' Despite the conflicts and tension, western economic growth was strong, and there was growing pressure on energy resources. China started to emerge as a major global economy and source of energy demand. The threats to the global climate were becoming clear to policy-makers.
- The decade 2010–2020 started with a severe global economic shock following the collapse of Lehman Brothers in the US. The banking bailouts were expensive, and austerity took hold in many major economies. We posit that public R&D, already a shadow of its former self, now hunkered down into defensive conservatism. Private capital was still available, but orthodox returns on investment were poor. Those who were 50 years old in 2015 had been amongst the youngest to see Neil Armstrong and Buzz Aldrin walk on the Moon in 1969. They had expected a world of bold, scientific endeavour based on large machines such as space-ships and fusion power plants. Many of this generation had become rich in the innovations that did occur in bio-medicine, mobile communications, software, and e-commerce. This reality extended below the famous billionaires now known for pioneering commercial space flight, but it also includes millions of millionaires curious about bold initiatives that somehow did not happen in the period 1990–2020.

We posit that there is now money available ready to absorb the extreme risks of a fusion energy proposition. We note the 2020 coronavirus pandemic and agree it will shape thinking for years to come, but we hope that rather than heralding a major economic collapse, it will merely be another future concern alongside climate change, terrorism, and cybersecurity. Fundamentally the new bold climate for innovation will be maintained. Noting the uncertainties facing the world, we expect that clean, safe, local and affordable fusion energy will remain a most attractive investment proposition for those with money, a desire to make a difference and an acceptance of risk. Noting the convergence of technology with an enthusiastic investment culture bodes well for fusion commercialization in the 2020s.

We close the book with an observation and a related question. Professor Ian Chapman of the UKAEA Culham Laboratory has described fusion energy research with the words: 'Fusion is probably the most collaborative field of science in the entire world' [1]. By way of illustration, he adds:

> It [ITER] is the most collaborative science project ever undertaken by humankind and it will demonstrate, we hope, that fusion really works on a commercial scale and can produce a lot more energy out than you have to put in to get the reaction going in the first place [1].

This culture of international openness and collaboration has been a major strength of the publicly-funded global fusion community. Indeed, the ITER experiment highlighted by Professor Chapman first arose as a result of a high-level political agreement between the leaders of the two superpowers of the Cold War. President Ronald Reagan of the United States and Premier Mikhael Gorbachev of the Soviet Union initiated the ITER programme at the Geneva Superpower Summit held in November 1985 [2]. If the USA and USSR could work together on high technology nuclear research and in this way nucleate the global collaboration we see today, then fusion has found a central place in the history of scientific co-operation.

Thirty-five years after the Geneva summit, we are now far closer to commercial development. The spirit of open and generous collaboration (with, for example, completely open intellectual property availability) is now starting to transition to the more competitive landscape of the commercial energy business. The start-up companies of the early 21st century are identifying key aspects of intellectual property, such as in high-temperature superconductors or in digital system monitoring and control. They are keen to realise the value of these innovations outside the fusion sector. Fusion needs these high technology innovations but it will not be the only sector needing these innovations and with a willingness to pay for them.

Finally, as we have seen in this book, the fossil fuel energy industry in transition is truly waking up to the potential of fusion with a sense of purpose not seen previously. We note the comments of the energy industry, including [3]:

- 'We see fusion technology as a promising, low-carbon future energy source,' Barbara Burger, president of Chevron Technology Ventures, 2020
- '(We are investing in fusion and a fusion start-up) because we believe in the technology and the company, and we remain committed to providing energy to the world, now and in a low-carbon future,' Sophie Hildebrand, Equinor's Chief Technology Officer, 2018

This book focuses on the fusion energy research and development sector in transition. That transition sits in the broader context of an energy transition where energy experts now see the beneficial attributes of fusion. For fusion, a key observation is perhaps that the greatest opportunities come from fixing other people's problems. Decarbonizing mobility, heat, and IT services represent a global challenge that provides nothing but opportunity for those seeking the commercialization of fusion energy.

References

[1] Lamont A 2020 *BP Technology Outlook Nuclear Fusion – Technology Special* (London: BP) (https://bp.com/en/global/corporate/what-we-do/technology-at-bp/technology-outlook.html)

[2] ITER Organization 2020 *The ITER History* (France: St. Paul-lez-Durance) (https://iter.org/proj/ITERHistory)

[3] Deign J 2020 Why are oil and gas companies investing in nuclear fusion? (https://greentechmedia.com/articles/read/why-are-oil-and-gas-companies-investing-in-nuclear-fusion)

Index